THE POWER OF ONE

The Role of the Individual in Nonprofit Organizations

Andon Rrapokushi

10 years of experience in establishing and supporting non-profit organizations

Original title:
The Power of One. ROLI I INDIVIDIT NË OJF
Translator: E. Bicolli

ISBN: 978-1-949597-37-0

Terra Nova Reports
Cold Spring, NY

In memory of Glenn and Martha Zimmerly,
Willard Heatwole and Paul Kropf

CONTENTS

FOREWORD

by Astrit Balliu

Ten years of experience of Andon Rrapokushi in the establishment and support of non-profit organizations, which mission was to support rural communities and rural development as a whole, are summarized in his book "The Power of One: Role of Individual in NPO". In point of fact, the author has been the protagonist of numerous efforts to establish Farmer Associations in several villages of Myzeqe and then to organize them in the Farmers' Federation of Myzeqe. The latter continues to be the most serious initiative of organizing Albanian agricultural producers in an organization and was intended not only to bring economic benefits to its members, but at the same time advocating the interests of small agricultural producers at the local and central government bodies.

The author is very realistic in describing the history of the establishment of Farmer Associations in Myzeqe area in Lushnja and the motives for their establishment. As usual in Albania during those years, even in the former Gradishtë Commune, the establishment of farmer organizations was the result of encouragement and direct support of foreign donors and the establishment motives were mostly social, or related to the existence of an organization as a condition for fundraising for infrastructure improvement (roads, water supply, canals) in the village. There have been many foreign missionaries devoted for years to raising the awareness of rural communities about the need to organize in associations, directly supporting their establishment. The author cites many of them in different episodes and human relationships, the book itself is dedicated to four of them, nostalgic for the years of collaboration, grateful for their human values and professional contributions, but still realistic and trustworthy to the reader.

Many believe and preach that the hesitation of the Albanian farmers to join agricultural producer organizations is a consequence of the bitter experience with former communist co-operatives. It is so true, that it tries to find all the possible names, just to not name them as co-operatives. It is important to note that despite calling themselves "associations", firstly "Fergor", "Fitorja" and "Gjelbërimi" (F1, F2 and F3 in the book), and the other new ones included later in the Farmers' Federation of Myzeqe, otherwise this commonly used slogan, realized the simple fact that the agricultural producers would join them if they would see and benefit economically from the membership in the association, would be real actors in the important decision making and if its funds were planned and distributed transparently.

The author has the merit of presenting truthfully the process of transformation of the above organizations from merely donator fund consumers into market-oriented producer organizations. Through multiple episodes that reflect the human relationships during this transformation, not rarely associated with tensions, and full of details on their economic progress for many years, the author faithfully reflects the efforts of various organizations to draft and then implement the business plans and to divide the functions and competencies of the "association's president" by those of the "economic activity manager". It is interesting to understand along the pages of this book the complexity of human relationships and the different views accompanying this process.

It has been almost a decade since the events described in this book occurred and it is good to see that many business practices, common today for the Albanian vegetable producers had their beginnings exactly at the economic activity of Farmers' Associations in Gradishtë (vegetable collection point, improvement of trading manner through the use of cardboard boxes, individualized product logos, etc.). In addition, the initiatives of these associations to provide their members with specific market information services (market information network) and to improve the producing technologies (extension service coupon system) remain innovative.

Rightly so, the author emphasizes the role of certain individuals in this ten years history of some of the most successful farmer organizations in Albania, grateful for their contribution and the progressive role and heartbroken in all the cases where the prejudices, or the changes in the vision for the prospect of associations' development have become a barrier to faster growth and organizational and financial consolidation. Not just them, local leaders. The different vision of various donors who supported the establishment and performance of Farmer Associations in Gradishtë created also problems in the associations' performance during the years of donor support and, moreover, the assurance of their survival after donors left the associations. And it is not just a matter of new economic activities proved unsuccessful. Not wanting to admit it, the author does not hide his "concern" for the sustainability of the associations' activity and their Federation after the program termination, while listing a large number of actions to be undertaken.

To conclude, though the first book and almost entirely based on facts and personal memories, Andon Rrapokushi's book *The Power of One* is an important contribution to documenting the establishment and transformation of Farmer Associations in Albania, the commitment and often the sacrifices of many dedicated people, Albanian and foreigners, who believed that beyond many failures favored by a constant climate of distrust in this process, the cooperation between small producers remains their best economic alternative under free market conditions.

Astrit Balliu, professor

Agricultural University of Tirana

INTRODUCTION

This book aims to share 10 years of working experience in rural development with the Albanian Mennonite Mission Foundation of (AMMF), several Non-Profit Organizations and Farmer Associations in Lushnja district during 1999-2008. This decade of work shows that leaders of these organizations have played a key role in the implementation of rural development projects in a part of Myzeqe region, confirming the fact that this important role of the individual and leadership is crucial to the functioning and performance of their organizations. The success or failure of a NPO or co-operative is inseparably linked to individual skills, character, level and vision of their leadership, which also guarantees other important factors within the organization, management, discipline, technology, cooperation, etc.

The events and facts described, have occurred in different places, but mainly took place in Lushnja district, where Divjaka Municipality is located also, which compose the main regions in Albania for the production of agricultural and livestock products and with the largest area of greenhouses in the country, nearly 600 ha. In this district with a population of around 145,000 inhabitants, exercise their activity some of the largest agribusinesses in processing of these products, trading agricultural inputs and selling agricultural and livestock products. The Fruit and Vegetable Wholesale Market is considered the second largest after Tirana in terms of volume and variety of products. The most powerful collection and exporting points for fresh fruits and vegetables, such as Doni Fruits Albania, Elian,

Tomato.al, ZDRAVA etc. are placed in Lushnja. To be mentioned are also Agro F&R Janko, with long experience in producing and exporting open field products, Lisar Seedlings in Divjaka (a big well-known seedling company), Agro Divjaka, well-known for producing and exporting open field products, and a unique farm for producing exotic products. Alba Seed in Çerma village is another big company for agricultural inputs and testing new vegetable varieties, as well as Agro LEV2. The most well-known factories in Albania for dairy products, Lufra and Ajka, are located in Lushnja district, which, otherwise is called as the green basket of Albania.

In this region of Myzeqe, which includes Lushnja and Divjaka, there are about 20 farmer associations and cooperatives, some of which have become known for their economic and social activity. Let's bring as an example the Agricultural Cooperation Society (ACS) Myzeqeja Farm in the village of Lumth, known for its long activity in the field of milk collection, and also for the gender equality. Led with competence and passion for many years by its two main leaders, Piro Jongari and Elvira Kola, Myzeqeja Farm is a concrete example that shows the effectiveness of a real leadership for agricultural and livestock cooperatives.

The Mennonite Foundation has the merit of giving the first spark to promoting cooperation between small farmers and their organization into associations and cooperatives in the Myzeqe region, and not just that; this foundation in partnership with other serious organizations has drafted and implemented a complex program of rural community development, as will be seen in this book.

Acknowledgments

Acknowledgment for the support of Professor Astrit Balliu, who wrote the foreword to this book and encouraged me by arguing the importance of such books that contain the experience accumulated over the years, during the implementation of rural development projects. A thank you also to Fatbardh Rustemi, writer, who gave his contribution to the book structure and appreciated the events and the information it conveys to the reader. Many thanks also to those who contributed to the publication of the book in Albanian language, Milton Heatwole, Melba Heatwole, Willy and Barbara Keener, and gratitude to Allen Umble, one of my best colleagues of the Albanian Mennonite Mission Foundation, who has been showing a continuous interest, encouraging me throughout the process of writing the book from the beginning.

Abbreviations

FFM – Farmers' Federation of Myzeqe

AMMF –Albanian Mennonite Mission Foundation

ICCO – Interchurch Cooperative for Development Cooperation

CERAI – Centro de Estudios Rurales y de Agricultura Internacional

AECID – Spanish Agency of International Cooperation for Development

RSC –Rural Studies Center

MCC – Mennonite Central Committee

VMBM – Virginia Mennonite Board of Missions

EMM – Eastern Mennonite Missions

RAAC – Regional Agricultural Advisory Center

OAA – Organic Agriculture Association

NPO – Non-profit organization

CTS – Car and Tractor Station

AFE –Agricultural Federation of Elbasan

ACS – Agricultural Cooperation Society

1. THE ALBANIAN MENNONITE MISSION FOUNDATION

It was July 27th 2004, when I received a SMS while traveling to Kosovo with a group of farmers from Lushnja, divided into two cars. The road to Kukës was difficult, mountainous and full of turns, and there was still time to arrive in Prizren (the Nation Road was not yet built). The SMS was sent by our lawyer from Tirana, Alfred Mustafa, who wrote: "Dear A. I notify you that the registration of the Federation has just been fully completed. Congratulations!". It was truly a great news for the entire group, consisting of leaders of three farmer associations, Fergor, Fitorja and Gjelbërimi (which in this book will be symbolically referred to as F1, F2 and F3), three representatives of the Albanian Mennonite Mission Foundation, two of whom were American citizens, Willard Heatwole and Daniel Hess, who had given a huge contribution to the process of establishing the Farmers' Federation of Myzeqe, as well as the director of the new wholesale fruit and vegetable market in Lushnja, Dritan Kolleshi.

The Albanian Mennonite Mission Foundation (AMMF) had made a valuable contribution to the entire area of Gradishte Commune at that time, assisting with a concrete and complex program of aid in the field of agriculture and livestock, which included the RLF-Revolving Loan Fund, agricultural and veterinary extension services, and other infrastructure development projects of the rural community. Providing free technical assistance to farmers and giving of over 70 soft loans, of which about 65 for small farmers of a mere interest rate of 6% -

8% in the villages Goriçaj, Këmishtaj, Mërtish etc., during 1999-2005, became a significant incentive factor for the employment of residents of this area, where the construction of solar greenhouses and the purchase of dairy cows had a huge impact, positively influencing the strengthening of small agricultural economies, the return of attention to land cultivation and hence the hope to create a better living in the village. Some soft loans with a low interest rate of 5% were also granted to farmer associations of the former Gradishtë Commune for the purchase of agricultural inputs by their members, and some years later also to the ACS of Myzeqeja Farm for the purchase of stainless milk cans, while neither banks nor other lending institutions in Albania provided and did not trust to farmer associations. In addition, 2 relatively large loans were granted to the Farmers' Federation of Myzeqe, the first to purchase 4,000 m^2 of land and the second to provide financial support for the construction of a large warehouse for the collection and processing of agricultural products, most of which was funded by the Spanish project, as will be seen later in this book. The positive impact of the work of the Mennonite Foundation exceeded the geographical boundaries, spreading throughout the Myzeqe area.

The man who had the vision and worked tirelessly for drafting and implementing the complex program of rural community development was Glenn Zimmerly, an American missionary who together with his wife Martha worked passionately and fondly for six years in Këmishtaj-Goriçaj area, earning the deserved respect of the peasants of Myzeqe area, a brilliant example that highlights the important role of the individual who inspired this change, with the unwavering support of the entire Board of Directors of the Mennonite Foundation *(photo no. 1)*. Glenn, with his prominent personality, with the love and dedication to help poor people and being a devoted and respected believer from all the inhabitants of the Myzeqe area, regardless of religion, provided support and funds from individuals, churches, organizations and private institutions in Ohio, Virginia and Pennsylvania. On the other hand, sending funds from individuals and American organizations to poor Albanian families in Myzeqe area is an evidence of solidarity coming from the distant continent to the simple and poor people, of love and truthfulness in helping people in need of a better life. With these grants the AMMF created the fund for micro loans.

Every farmer who wanted to receive a micro loan for a project on his farm filled out a simple form with the following information:

ALBANIAN MENNONITE MISSION FOUNDATION

1. Personal data:

*Name, Fatherhood, Surname*____________________________

Date of birth ____________ *Place of birth* ________________

Residence ____________ *No. of passport* ________________

Name of spouse ______________ *No. of passport* ____________

*2. The purpose of obtaining a loan*_________________________

____

____

____

3. The place where business will develop ____________________

____

4. Guarantees provided by the client (building, cows, land, etc.)

____

5. The amount of loan he wants to take ____________________

6. When does he intend to repay it? ________________________

____

Date of form completion __________________

Applicant

(name, surname, signature)

THE AGREEMENT

Upon the receipt of loan by the applicant, it must be clear that the loan repayment will be made in accordance with the contract, which he signs with his free will. The installment payment date will be respected under the contract and no excuses will be accepted by the borrower. The interest rate on the loan is calculated at 8%. It is accepted that all the expenses and income to be recorded in an account book that can be checked at any time. The Project implementation will be reviewed several times until the loan is effective.

(name, surname, signature)

This was the content of the form completed by the farmer and in case of approval of the request, it was drafted a bilateral contract before the notary and for each borrower a special file was created. About 7-8 days before the payment of each installment, to the borrower was sent a notification with this content:

Mr. ______________________________

Dear borrower!

In accordance with the contract, the date for the payment of your loan is approaching, and in this case we remind you of the promise made to us prior taking the loan: you will be very correct with the repayment of the loan.

Please read the contract you signed in the notary office and you will see that on date ______ you have to pay: ________________ ALL.

We will be visiting you on ___________.

Remember that there are other families who need a loan, who are your friends or cousins, or your neighbors. Our program is for a joint progress.

Thank you for understanding!

Whenever different groups, either from Albania or from America, went to the area where the community development program was implemented and asked the residents, farmers, men and women, gave the highest appreciation for the work of the Mennonite Foundation and all its employees, and in particular for Glenn and Martha Zimmerly, in whose honor they have also placed a marble memorial plaque at the center of Goriçaj village *(photo no. 2)*. The Council of Gradishtë Commune on March 27th, 2002, by decision no. 37[1] awarded to Zimmerlys the title "Honorary Citizen of the Commune". The four motivation points of this decision state:

For the excellent work they have done as members of the Albanian Mennonite Foundation in our commune for 6 years.

1. In terms of granting loans for the construction of greenhouses of farmers.

[1] Decision no. 37, dt. 27.03.2002 Council of Gradishtë Commune

2. In terms of providing contemporary knowledge and technical assistance to our farmers through the winter school.

3. As initiators for the establishment of associations and key managers for finding donors who have invested in the commune.

In Glenn's absence, this title was handed over during a festive ceremony, to a distinguished American missionary, Willard Heatwole, from the mayor of Gradishtë, Mr. Vasil Sema. Williard, receiving the title said modestly: *"I cannot wear Glenn's shoes, they are too big for me, but for me is a honour and respect to receive in my hands this title you have awarded to him"*. Glenn's dream was to work for many more years in Goriçaj and see this village developed with an advanced agriculture and livestock, where farmers cooperate with each other as members of their co-operative. He left for the United States because of diabetes but his vision was becoming reality through the correct implementation of the social and economic development projects of AMMF in two aspects: theoretical-practical activities and training, and financial support for the socio-economic needs of the community and the poorer farming families.

Dedication to the work that started in the Lushnja district by the Zimmerly couple continued for several years by other employees of the Mennonite Foundation, Willy and Barbara Keener, Daniel and Mary Hess, Sonya Harnish, Deb Kilheffer, Steve Campbell, Dan and Melike Smenge and the Heatwoles – Willard and Melba. Willard Heatwole, who passed away in October 2015, was a man of great love for ordinary people, besides being a devout Christian believer, equipped with profound technical knowledge and an expert of rural community development, a prominent and active individual in many activities of farmer associations and of the

Farmers Federation of Myzeqe, who provided valuable advice on the organization of work to small farmers. In several meetings on the problem of water salinity used for irrigation of greenhouses in the village of Goriçaj, since 2003 he warned farmers about this problem, which could deteriorate in the next years, increasing their level of awareness with arguments. Willard took as a concrete example the water from the village vein, which was the main source of water for the greenhouses, where some families threw various plastic waste or emptied into its banks organic fertilizer wastes, creating water pollution, instead of protecting it as a natural gem that could be used not only for irrigation of greenhouses but also for other purposes for the benefit of the village.

There was also some concern about the water coming from Thana reservoir. Since the Myzeqe area was receiving the irrigation water from this large reservoir, on Willard's initiative, in August 2004 we initiated a trip along the Devoll River, from which this reservoir was supplied, one of the largest in Albania and Balkans, with capacity 60 million m^3 of water. The purpose of the trip was to know the quality of water and to determine the cause and place where the increase of the salinity level began. Together with Willard and one of my sons, we passed along Devoll's flow from Gramsh, across the rural road to Maliq, between the mountains and with a very beautiful landscape, and then completed the trip to Vlashuk canal from where the water entered the reservoir. The water samples taken in many different places, were all analyzed by evaluating the main elements, especially the percentage of salt, and were reported to farmer associations and farmers directly at the meetings organized for this purpose. Willard's son, Conrad Heatwole, professor at Virginia Tech University, during his visit to Albania in 2003, drafted a project idea specifically for the problem of water salinity in Lushnja district, which he presented together with me at USAID offices in Tirana. One of the main objectives of this document, which included 3 timeframes, was to provide farmers with a clearer understanding of the sources and nature of water salinity in the Myzeqeja area and of the irrigation management.

A valuable contribution of the farmers' honorable advisor, Willard Heatwole, was also raising their awareness on the role of atmospheric and climatic conditions in agricultural products. He established links with the Meteorology Institute in Tirana, through which the agro-meteorological weekly bulletin was taken and distributed to farmers on a regular basis through the Farmers' Federation. At the meeting of September 10th, 2004 at this Institute, together with Willard we had a conversation with the head of the Agro Meteorological Department, the well-known expert Prof. Petrit Zorba for a bilateral cooperation and completed the annual registration to provide regular scientific information on the agro-meteorological situation to the FFM – Farmers' Federation, which was newly established. This information was sent to us regularly by email once a week in pdf format. It was 2004, when there were no smartphones and iPhones, and in our country in Albania, especially in rural areas, there were not the nowadays opportunities for providing these data to farmers.

In addition to contributing to agriculture, Willard Heatwole provided a valuable contribution to the provision of running water in the primary school of Spolata village, a hilly area far from the main road. The headmaster of the school, Hëna, had addressed the problem of the lack of water at the school to the relevant institutions, a problem that had not been resolved for a long time, but Willard's involvement in this issue caused this problem to be solved well and very quickly. He found a way of cooperation with the local government and with the school directorate and after 2-3 meetings with the mayor of the Gradishte Commune, Vasil Sema, it was prepared a rehabilitation project for the surroundings of the school by drilling a well, for which, CAFOD organization contributed also. Willard drafted the project of the water distribution line from the well inside the school, and in addition to the municipality's contribution, contributed personally to the purchase of a small amount of hydraulic materials and followed the implementation of this project.

2. THE F1 ASSOCIATION

2.1 Establishment and organization of the association

The establishment of the first Farmer Association in Lushnja district in June 2000 by the farmers of Goriçaj, was a great achievement in this area. It started with an initiator group of 6 farmers from the village of Goriçaj who, strongly encouraged and supported by the Albanian Mennonite Mission Foundation, worked for several months until its establishment and the founding meeting was held on June 5th, 2000, in which 54 farmers became members and a board of seven members was elected. The registration of this association in court took place on 27 September 2000. The process of establishing the F1 association followed the way, which is clearly described in the book *"Farmer Co-operatives and their organization"* by the professors of the Agricultural University of Tirana Bahri Musabelliu and Maksim Meço, who in the chapter *"Organization of a Farmer Co-operative"* explain the stages and moments through which this process passes, the importance of leadership and consulting, and the key elements of success[2]. The main external factor that guaranteed the development of this process was the AMMF, a foundation that seriously engaged in the socio-economic development of the rural community in Goriçaj and which

---

[2] "Farmer Cooperatives and their organization" pg. 65, B. Musabelliu, M. Meço, Tiranë 2003

confirms the important role of consultancy emphasized in the aforementioned book.

The F1 Annual General Meetings during the first 6 years, 2000-2005, were a model of such organizations with a well-designed agenda, which was distributed in advance to the members, with an annual report analyzing the activity of the association and with a report of the governing board of the association. A voting commission was nominated for the elections, which conducted the secret ballot for the governing bodies of the association, counted the votes and announced the new governing board, which in turn elected the president, normally the one who won the majority of votes *(photo no. 3, no. 4)*. The assembly, according to the statute, held the general meeting once a year, but in Goriçaj it met more frequently. Let's see for example the elections held at the end of November 2004 for board members, which were won by a respectable woman and skilful farmer, Mrs. Ariste Ziu with 18 votes in favor, while the other 4 members received respectively 14, 13, 11, and 11 votes. According to the statute, the one who won the majority of votes became president of the association and from the other members of the board was elected the vice president and the secretary of the association. However, Ariste did not accept to be president and thus the presidency passed to the second one, Andrea Thomaj with 14 votes, a very good farmer with long experience in cultivating vegetables and in the organizational field, while Mrs. R. Billo, who won 13 votes, was elected secretary. (At this meeting, the assembly decided that the number of board members to drop from 7 to 5). The election of two women at the board of the F1 association was an innovation and a positive example for increasing the role of women in the governing bodies of the farmer associations and in the rural community.

So the change of the first F1 president was done democratically, with transparent elections, a phenomenon that does not happen often with farmer associations in Albania, where the president rarely changes. The leader of the association in question led the association for the first five years, being elected twice during this period, and on the occasion of him leaving the position, the Mennonite Foundation in collaboration with the F1 Association organized in December 2004 an activity at Mondial Club in Goriçaj, where a large number of members of the association and representatives from the AMMF participated, on behalf of which Sonya Harnish greeted the F1 association and praised the work of its first president during the implementation of the first ICCO project, the story of which is described below.

2.2 Projects for socio-economic development of the village

The F1 Association, having a very active board, although newly established, but also with the strong support of the Mennonite Foundation, in December 2000 managed to win a package of projects for its community under the "Rural Community Development Program" , co-funded by ICCO Netherlands, in the amount of USD 271,478. We cannot but mention Mrs. Pien de Lange, a Dutch extraordinary woman, full of energy and kindness, who came from Holland in Goriçaj in 2000, to see closely the socio-economic problems of the village and to meet directly with the board of the association. So there are two prominent figures at the center of the development project of this rural community: Glenn Zimmerly of the Mennonite Foundation and Pien de Lange of ICCO. These were two people of great heart, who believed till the end in the peasants of Gorіçaj, and that the funding would not go in vain, on the contrary, would improve their lives, just as it did. The persistence of Glenn and Mennonite Foundation and the arrival of Mrs. Pien de Lange (known for her contribution in many districts of Albania as one of ICCO's main figures), culminated in the signing of the project by both parties, ICCO and F1, on 13 December 2000. Two weeks later, on December 28th, another important agreement was signed by four organizations: AMMF, ICCO, F1, and Fier RAAC. *(See Annex 1)*. This Operating Agreement with four articles sanctioned in the first article the fact that the F1 association would take over two major projects, respectively, the water supply construction and the rehabilitation of village roads, as well as the telephone communication system; article 2 sanctioned that the Mennonite Foundation would be responsible for managing three projects in the field of agriculture, respectively the winter school, the agricultural counseling program and the poultry project. Article 3 defined that the project office would be under the management of Willy Keener, Executive Director of the Mennonite Foundation, and the Project Manager. While article 4 defined as responsible for the implementation of the winter school and the agricultural counseling contract, the Fier Regional Agricultural Advisory Center (RAAC or AIC, Agribusiness and Information Center) with Director, Mrs. Zh. Zylalaj. After being signed by the main representatives of the three Albanian organizations, the agreement was sent by fax to Netherlands to be signed by ICCO, which as a project funding organization was very interested in their success. Mrs. Pien, on 11 January 2001, sent an email to me confirming the Operating Agreement on behalf of ICCO *(see Annex 2).*

Very constructive cooperation and correct relations with ICCO continued at the same level as with Pien de Lange, even with Gre Kornelius, with Joost Van den Hee, etc. The high degree of engagement of the rural community and other actors involved in the project during its implementation, as well as micro loans program focused on Goriçaj, transformed completely the village; there were significant socio-economic changes, improving significantly the lives of its inhabitants. A 7 km long water supply was constructed together with the internal network, bringing water to each house, and there were rehabilitated the roads full of holes of the village, while the entrance road to the village was asphalted *(photo no. 5).*

For the realization of these two main projects, the water supply and the village roads, the F1 Association launched an open tender, attended by 4 companies from Lushnja and Fier districts. This tender was held at the office of the association and a commission composed of members of association's board and representatives from the Mennonite Foundation, announced the winner Albstar Fier with President Mr. Sadik Ismailaj. The works were carried out with a very good quality. While the well project was realized by the engineer Prof. dr. Romeo Eftimiu of ITA Consult-Tirana Company and the drilling of the water supply well at 180-ml depth (*from 200 ml planned – see annex 9 "Well Project Contract"*) in source 7 km far from the village was carried out by Seman sh.p.k. The implementation of these projects was monitored by the talented engineer Mr. Vlashi Bano, who worked hard every day and presented relevant reports both at the F1 association and the Mennonite Foundation.

The F1 Association, although newly established, in 2001 managed to contribute to the potable water project by $5,860 (or about 838,000 ALL according to the exchange rate at that time) which was provided by collecting a fixed fee from180 families of Goriçaj village. For this the merit goes to the board of the F1 association, led by its first president. The implementation of this project and the road rehabilitation process also led to the opening of many jobs for the village residents, who were hired by the implementation company, Albstar. In September 2001, these projects were successfully completed. The inauguration ceremony that was organized in the village of Goriçaj was attended by hundreds of inhabitants of the area, as well as by the parliamentarian of the area Mrs. Arta Dade and the Executive Director of the Mennonite Foundation, Mr. Daniel Hess, who cut the inauguration ribbon together with Sonila Prifti, the F1 representative *(photo no. 6).*

2.3 Projects with agricultural profile

In addition to the two major village infrastructure projects, several important agricultural projects were also implemented:

- The Poultry Project, which aimed at improving the food structure and increasing the incomes of poor farmer families.

- The Winter School of farmers, with two main objectives:

a) To help farmers with a qualified scientific opinion in the field of agriculture and animals breed improvement.

b) To increase the level of farmers to produce qualitative agricultural and livestock products.

- The Agricultural Advisory Service, with the objective: to bring progressive opinions by agriculture specialists to the farmers throughout the year (*photo no. 7*).

- The Communication System, with the objective of improving telephone communication (at that time in Goriçaj there was no telephone, neither fixed nor mobile, and the communication needs were too great).

- The Association Office and Service Center, with the objective of consolidating the F1 Association and facilitating the services to its members.

Meanwhile, for the realization of two projects in the field of agriculture, the AMMF and the F1 Association signed an agreement with RAAC in Fier, a Dutch model which was implemented even in Durres. Among the objectives of this center was stimulation and cooperation among farmers, as well as providing market information to farmers and agro-processors. The two projects for which the agreement was reached were the Winter School with farmers and the Advisory Service. Fier RAAC Center, with advisor the well-known Dutch development expert Gerrit Holtland (who a few years later was also an external consultant of the Farmers Federation in Këmishtaj) made available his qualified staff, Zhaneta, Vasilika, Fredi, Gramoz, veterinarian Todi Dano and others *(see annex 3)*. In addition to these two projects, the AMMF and the F1 association successfully implemented several times the poultry project during 1999-2002.

The Winter School, initiated by Glenn, which was implemented for four consecutive years during the winter period of 1999-2003, was in fact an authentic training program where the farmers of Goriçaj and other surrounding communities attended and debated on various topics about production technologies in greenhouses, the hygiene and maintenance of dairy cows, seeds and agricultural inputs etc., topics which were presented by qualified specialists *(photo no.8)*. About the manner of organization of the school and the problems to be addressed there, on 26 October 1999 was organized an extensive discussion with the farmers of Goriçaj, who were divided into three teams, each choosing a farmer as team leader, and each team putting forward its own ideas and the socio-economic problems of the village. Part of these teams was Glenn (in the first team) and Willy Keener (in the second team).

Among the first specialists invited to the Winter School, professor Hysen Cobani presented to farmers of Goriçaj the new varieties of vegetables and the activity of TARADIA Company and the products it traded, having the exclusivity of the Dutch company "Sluis & Groot", while the Executive Director of Lisar Seedling Company, Josif Gorreja, held several presentations on the new varieties of vegetables in the greenhouses and the chemical fertilizers that this company traded. Part of the topics of the Winter School were the trainings of the farmers' leaders too.

In addition to the theoretical topics, the Winter School project included working visits of farmers within Albania, such as in the "Light Force" livestock farm in Lezha, in the Floryhen poultry company in Durrës, in agribusiness centers in Fier, etc., and in neighboring countries, in Macedonia (Strumica) and Kosovo. The Winter School was open from December to March or January to April, while the Advisory Service included 2 periods of the year, referring to 2 greenhouse seasons, respectively February-June and August-November.

Though at the end of 2000, for health reasons, Glenn left Albania together with Martha, he sent from America some video for the Winter School with farmers, which were used for training and they continued to be interested in this village and its residents until the end of his life. The Winter School took place at "Mondial" Bar in the village center. In order to make it more efficient, and to improve the services to farmers, two other small projects contributed, "The Association Office and Service Center", and "The Telephone Communication System". In February 2001, the board of the F1 association decided to buy five mobile phones, two electric heaters, a VCR Multi System video camera, a generator to use when there was no electricity and office equipment for the association. These devices were purchased in Tirana and helped to implement in better conditions the trainings and the community development projects in Goriçaj.

As emphasized above, all members of the F1 first governing board have been active, cooperative and dedicated and I would like to distinguish Mikail Davidhi, an example of a leader who with correctness and initiative supported the new changes, a hard working farmer, and also an extraordinary family, man and father of two excellent children, with the utmost care to well educate them, who unfortunately departed from life very soon.

Constantly committed to work was the financial administrator of Goriçaj association, Llazar Stavri, who kept very orderly documentation in his office, who was disciplined and correct and always willing to cooperate with colleagues and farmers. He is an example of what qualities and abilities should have those who work with farmer associations and co-operatives.

The F1 association's experience in the field of management and implementation of rural development projects and its close cooperation with the Mennonite Foundation, was subject of a TV show on TV 4+ in April 2002. This experience was extended to two other associations (F2 and F3), which in 2004, together with the F1 association, established the Farmers' Federation of Myzeqe. Below is shown the model used for the Winter School calendar, implemented in these three associations in 2003:

Calendar of Winter School No. 4, year 2003

MONTH	JANUARY	FEBRUARY	MARCH	APRIL
No. of activities	**4**	**5**	**3**	**1**
Date and type of activity	d.10, training for greenhouses (in Goriçaj)	d.7, training for potatos (in Fierseman)	d.14, training with the leaders –F1 association governing board	Working visit with a group of farmers, members of three associations in Kosovo
	d.17, training for viticulture (in Fierseman)	d. 7, training for greenhouses (in Mërtish)	d. 21, training with the leaders – F2 association governing board	
	d. 24, training for greenhouses (në Mërtish)	d. 14, training for livestock (in Goriçaj)	d. 28, training with the leaders - F3 Association Administrative Council	
	d. 31, training for greenhouses (in Goriçaj)	d. 21, training for vegetables (in Fierseman)		
		d. 28, training for livestock (in Mertish)		

The data below represent the calendar of agricultural consultations implemented in six villages, from February to November 2003:

Calendar of Agricultural Extension Program No. 4

Collaboration between Mennonite Foundation, Fier Agricultural Advisory Center and Farmer Associations F1, F2 and F3

GORIÇAJ - KËMISHTAJ - MËRTISH - FIERSEMAN - GRADISHTË - SPOLATË

February - November 2003

FIRST PHASE, 22 activities					
Month	**February**	**March**	**April**	**May**	**June**
No. of visits	4	6	7	3	2
Date and type of visit	d.10, in greenhouses	d.5, in greenhouses	d.1, in greenhouses	d.2, in greenhouses	d.3, in vineyard, fruit trees
	d.11, in greenhouses	d.11, in vineyard	d.3, in livestock	d.14, in vineyard	12, in vineyard
	d.25, in greenhouses	d.13, in greenhouses	d.4, in greenhouses	d.22, in livestock	-
	d.26, in greenhouses	d.18, in greenhouses	8, in vineyard and fruit trees	-	-
	-	d.21, in greenhouses	11, in greenhouses	-	-
	-	25, in open field vegetables	15, in open field vegetables	-	-
	-	-	22, in livestock	-	-

SECOND PHASE, 10 activities

Month	August	September	October	November
No. of visits	2	4	3	1
Date and type of visit	d.21, in greenhouses	d.2, in greenhouses	d.3, in greenhouses	d.7, in greenhouses
	d.22, in greenhouses	d.12, in greenhouses	d.10, in livestock	-
	-	d.16, in open field vegetables	d.17, in open field vegetables	-
	-	d. 26, in open field vegetables	-	-

Both of these agricultural projects were sponsored in 2003 by the MCC (Mennonite Central Committee), Bosnia and Herzegovina's office in Sarajevo headed by Randy Puljek. The MCC also contributed financially to two other community development projects: firstly in the rehabilitation of the rural road Goriçaj-Këmishtaj. The rehabilitated part was 770 m long in a miserable, very bad situation, with deep pits and mud, but which was especially necessary for the pupils walking every day from Goriçaj to the school of Këmishtaj. The entire length of the road to the school was over 2 km and the works were done during February-March 2003. The families living along this road provided a modest financial contribution, among which the Daniel Hess family who lived in Goriçaj. The MCC also sponsored the process of establishing the Farmers' Federation of Myzeqe, when this process was in the early stages, having again the Mennonite Foundation as a partner organization *(Annex 4)*.

3. THE PROJECT OF POULTRY AND VEGETABLE PROCESSING

3.1 Implementation of Poultry Project

This project was implemented not only in Goriçaj, but also in many other surrounding villages, including Pamper, Mërtish, Sopez, Krutje, etc. The first poultry project consisted in the purchase of 2040 - 24-hour small birds, of the Isa Brown breed, with a fast growth cycle of only two months using the appropriate concentrated food. Together with the Executive Director of AMMF, Willy Keener, at the beginning of August 1999 we went to Ioannina in Greece, and bought this quantity of birds, having previously obtained the relevant documents from the Veterinary Service and the Livestock Production Directorate in the Ministry of Agriculture and Food in Tirana.

In Kakavija Customs House, however, despite having all the necessary documents and passing Greek customs without any problems, we encountered some difficulties, moreover considering the fact that it was a sunny August day and in the car we had 2040 newborn birds. So we arrived late at night in Lushnje and immediately the whole group of the Foundation employees took care for feeding and sheltering the small birds, in order to not have any losses. The birds were sheltered in thc big hallway of the Smenges house and the next morning started their distribution to the farmers who had ordered them.

The second poultry project included the purchase of 2000 24-hour birds, which were purchased in early May 2000 also in Ioannina and as in the first case there were issued the import license and the veterinary certificate from the Ministry of Agriculture and Food and the district Directorate of Agriculture.

The third poultry project started a year later, in May 2001, at the demand for birds and food from the farmers of some villages in the Myzeqe area and on 30 May 2001, 3280 birds and 574 kg of concentrated food were again purchased in Greece *(photo no. 9 and no. 10)*.

The implementation of the poultry project was characterized by a very good cooperation with Hiqmet Driza, owner of a big poultry business in Patos, who kindly assisted the implementation this project several times, offering the appropriate food as well as small birds when there were more demands from the farmers.

The project provided farmers with free bird vaccination throughout their growing cycle. The Mennonite Foundation performed regular monitoring throughout the growing period of the birds by keeping accurate statistics on the number of birds at the beginning and at the end, the type of food used, the diseases, if any, and the average weight per herd; and many farmers managed to have poultry of around 2 kg after two months, as was foreseen for this breed.

The project covered:

- the cost of transport for the purchase of the birds and their delivery to every village.

- the purchase of 9,000 doses of vaccine, including 3 types of vaccines.
    - the cost of transport for the delivery of vaccines to every village where birds were bought.
- the price reimbursement with 5 ALL / bird.
- the cost of transport for the purchase of concentrated food in Durres and Patos.
- - the monitoring and technical assistance.

The monitoring and assistance took place at certain times, but let's see the data from the monitoring conducted two months after the purchase of birds in three poultry projects:

1) The project of 1999 begun on August 7th, with the purchase of 2040 24-hour birds; the monitoring took place on October from 6 to 11 in most of the farmer families who had bought birds.

Name of farmer	**Date of control**	**No. birds at the beginning**	**No. of bird s alive**	**Average weight**	**Recommended food**	**General situatio n**
T. Stambolliu	6 October	110	90	1,6 kg	Yes, Concentrated	Very good
Th. Hila	6 October	100	30	1,2 kg	No	Asthma
Ll. Gjata	6 October	50	44	0,7 kg	No	Good in general
M. Nushi	6 October	55	21	0,5 kg	No	-
K. Llupo	6 October	40	18	0,3 kg	No	-
W. Keener	7 October	26	24	1,7 kg	Yes	Very good
A. Kushi	7 October	15	15	1,8 kg	Yes	Very good
D. Baze	8 October	300	50	0,6 kg	No	Chicken-pox and asthma

A. Ziu	8 October	30	28	1,8 kg	Yes	Very good
S. Dila	8 October	50	12	0,6 kg	No	Chicken-pox
M. Tashi	8 October	20	18	1 kg	Yes	Good
J. Ziu	8 October	100	5	?	No	Grave chicken-pox
Al. Ziu	8 October	50	0	?	No	Grave chicken-pox
A. Hila	11 October	60	35	1 kg	Yes	Partly chicken-pox
N. Ciko	11 October	30	27	1,2 kg	Yes	Very good

The data clearly show that along with other conditions, the use of the recommended food is an important element to succeed. But this was the first project and we had no experience. While in the second poultry project were achieved better results.

2) The second project of 2000 began on May 11; the inspection took place after 2 months, from 12 to 14 July.

Name of farmer	**Date of control**	**No. birds at the beginning**	**No. of birds alive**	**Average weight**	**Recommended food**	**General situation**
M. Tashi	12 July	60	55	1 kg	At the beginning. Now corn	Very good
S. Dila	12 July	20	20	1,1 kg	At the beginning. Now wheat	Very good
J. Llogone	12 July	50	25	0,8 kg	Yes	Complaints on food quality

A. Llogone	12 July	30	28	1,1 kg	At the beginning. Now wheat	Very good
T. Stambolliu	12 July	50	38	1,1 kg	At the beginning. Now wheat	In general good
Todi Stambolliu	12 July	20	17	1,3 kg	At the beginning. Now wheat	Very good
V. Gjoka	12 July	10	9	1,1 kg	At the beginning. Now wheat	In general good
N. Leshi	12 July	15	10	0,9 kg	1 month of concentrated food	In general good
M. Leshi	12 July	15	14	0,9 kg	At the beginning. Now wheat	Good. Complaints on food
A. Ziu	12 July	50	50	1,2 kg	At the beginning. Now corn and wheat	Very good
Th. Prifti	12 July	40	40	0,5 kg	100 kg concentrated food	Small Complaints on food
Joshua Keener	12 July	16	10	1,6 kg	Yes	Good. Problem with feet
Moses Keener	12 July	10	6	1 kg	Yes	Good. Problem with feet
E. Lule	13 July	30	30	0,6 kg	At the beginning. Now wheat	Good but small
S. Ziu	13 July	80	80	0,8 kg	At the beginning. Now corn	Good. Problem with feet
S. Meço	13 July	15	12	0,8 kg	Corn and wheat	Good
P. Kola	13 July	100	70	0,5 kg	Corn and wheat	Too small

K. Mita	14 July	10	7	0,7 kg	At the beginning. Now corn	Good
A. Kushi	14 July	19	18	1,8 kg	Yes	Good. Problem with
Th. Mita	14 July	11	10	0,7 kg	At the beginning corn	Good. Complaints o

3) The third project of 2001 began on May 30; the last monitoring was done after 2 months, on July 26 and 27, just like the previous two times, by the AMMF team. It is the project with the best results.

Name of farmer	Date of control	No. birds at the beginning	No. of birds alive	Average weight	Recommended food	General situation
Trifon Stambolliu	26 July	60	60	2,1 kg	Yes, concentrated	Very good
Todi Stambolliu	26 July	30	29	1,9 kg	Yes	Very good
T. Lena	26 July	50	48	2 kg	Yes	Very good
J. Llogone	26 July	30	26	1,5 kg	Yes	Very good, clean
A. Ziu	26 July	35	35	1,4 kg	At the beginning. Now wheat, flour	Very good, clean
A. Gjoka	26 July	50	49	1,1 kg	At the beginning. Now wheat	Good, clean
D. Mita	26 July	20	20	1,1 kg	At the beginning. Now wheat, alfalfa	Good conditions, clean
S. Ziu	26 July	60	45 (of which 5	0,9 kg	Wheat	Very good, have

			are consumed)			consumed 5 for food
A. Llogone	27 July	70	64	1,1 kg	Little Concentrated Food. Now bran, wheat	Good, clean
L. Leshi	27 July	20	18	1 kg	At the beginning concentrated	Good, a little bit welded
M. Nushi	27 July	10	10	1,2 kg	Concentrated food	Very good
K. Çela	27 July	10	7	1,9 kg	Yes	Very good
Ll. Pina	27 July	60	60	1 kg	Concentrated, then wheat	Good conditions
I. Tafa	27 July	20	18	0,5 kg	No. Wheat	There is no special place for birds
Xh. Maka	27 July	10	10	1 kg	Yes, concentrated	In general good

The best results of all three projects were achieved in Goriçaj, with the heavier rooster reaching 2.5 kg after two months. In each village, all the farmers involved in this project bought birds mainly to meet their food needs. The only case for business purposes was that of a farmer from Mërtishi who bought 1500 birds in the third project.

By making a simple analysis of the situation at the beginning and at the end of the two months for the 3 phases of the poultry project, it results that from the data obtained during the monitoring, the results have improved and the loss in birds has decreased significantly year after year, with only 5.8% loss in the third year. For more specific information, refer to the table below:

Number of birds and losses during the monitoring phases, where visits were made:

Project I, year 1999 Monitoring 15 families			Project II, year 2000 Monitoring 20 families			Project III, year 2001 Monitoring 15 families		
Birds at the beginning	Birds at the end	Looses in birds	Birds at the beginning	Birds at the end	Looses in birds	Birds at the beginning	Birds at the end	Looses in birds
1036	417	619 or 59,7%	651	549	102 or 15,7%	535	504	31 or 5,8%

Conclusions:

A large number of farmer families benefited and learned from poultry project, which was repeated several times. The results can be summarized by three points:

- The conditions of poultry and the care for the small birds were improved, especially in the first two to three weeks.
- The importance of the type of food and clean water for feeding birds became clear.
- A good experience was created and farmer's knowledge was increased.

The poultry project was only one of the projects implemented by the AMMF, while at the same time were implemented other important projects as mentioned above, winter school, extension service etc. As indication of the deep respect and appreciation of ordinary people for the work of the Mennonite Foundation and especially of Glenn Zimmerly in Goriçaj, I would like to mention the words of a farmer woman, Donika Davidhi; when I went to her house with Sonya and Deb Kilheffer during the monitoring of the poultry project in June 2001; laughing she said: *"People here in Goriçaj would like to vote for Glenn as their representative in the Albanian parliament."*

3.2 Project of processing and conservation of fruits and vegetables

One of Glenn's plans in the framework of his vision for the development of rural community and improvement of the lives of farmer families was the project of processing and conservation of vegetables and fruit in Goriçaj. Glenn, as well as his wife Martha, had many technical and practical knowledge in this field, and he had collected many written materials, books, leaflets for simple processing and conservation equipment and concrete information for such small projects, but with a great impact in the improvement of life in the village. Such materials he collected especially from Dixie Canner Company, with which he had set up a correspondence to obtain accurate information on the technology, type and prices of household conservation equipment manufactured by Dixie. Pursuing the correspondence and in response to Glenn's letter, Dixie Canner Co. in Georgia, USA, sent a package of written materials for the equipments, accompanied by a letter written by the manager of the company, Mrs. C. Simmons *(see Annex 5)* who among other things wrote:

"Dear Mr. Zimmerly, thank you for your inquiry on Dixie equipment. Enclosed please find our latest brochure on Dixie equipment .... All equipment is priced with regular equipment as shown .... Please provide electrical requirements and shipping directions for a more accurate quotation ... ".[3]

---

[3] Dixie Canner Co. Cynthia J. Simmons, Office Manager, August 24, 1998 – Letter to Glenn Zimmerly, Lagjia Gafur Muço, Lushnje, Albania

To make this project more understandable to farmers in Goriçaj, Glenn brought an example from a small country comparable to Albania, Moldavia, which had a very good experience in the field of processing agricultural products under household conditions, and to illustrate it better Glenn referred to a project implemented in the village of Antonesti. Based on this project, Glenn's idea was to buy a set of equipment for the processing and conservation of vegetables, fruit and juice production. These equipment would be installed in the village and would be under the management and control of the F1 association to serve to all village families with their products. The site to be installed initially would be simple and thus a large empty container was provided in Durres. I went to the port of Durrës where a Benz trailer loaded the container which the Mennonite Foundation provided through a regular procedure from trucks that had come to Albania with aid for Kosovo people in the spring of 1999 and brought it to Goriçaj in the backyard of M. Nushi, at the disposal of the F1. In addition to the equipment, a training program was planned for women farmers for the processing and conservation of vegetables and fruit, with the assistance of the Ball Corporation company, specialized in this type of business.

When the package of Goriçaj rural community development projects was proposed to the Dutch organization ICCO for funding, the largest and the most important of them, such as water supply, road rehabilitation and small-scale agricultural profile projects, were approved, but not the one of fruit and vegetable conservation. With Glenn's departure due to his illness, no complete study was carried out for the realization of this project, which required qualified people and financial support. Meanwhile, the implementation of the approved projects, which required a maximum commitment from the AMMF and the F1 association, had begun.

By mid-December 2000, after a visit to the ABC medical clinic in Tirana, Glenn and Martha, according to the doctor's advice, were about to return to their home in Ohio, America. A few days later, as I returned with Martha from Këmishtaj, after a visit to a borrower's home, she said thoughtfully: "It will be very difficult for Glenn to leave Albania and the very good people he has known here". I told Martha that in the first 2-3 days of my employment by the Mennonite Foundation, coming from a completely different sector, education, I expressed to Glenn concern about the nature of the work I would do, but he immediately removed my discomfort, expressing his belief that I would do my job very well and we would work together...; these were words of great importance to me. I accompanied them to Rinas airport on December 29th, 2000.

Despite leaving, even from his home in Orrville, Ohio, Glenn did not give up on helping Goriçaj's community. The processing and conservation project was in his heart, he revised it in all the elements, making a detailed cost re-evaluation for equipment and machinery, for land acquisition, building, offices, as well as calculation of VAT, and almost a year later, at the end of October 2001, he sent me the project by email from Orrville. We could not find a donor to fund this project of $ 214,665, but I am convinced that if Glenn was still among us, he would also have implemented this project.

4. THE KOSOVO EXODUS

Leaving aside the description of the projects realized for the development of the community with the F1 association, let's focus to another important aspect of the activity of the Mennonite Foundation in Lushnja district, that of assisting thousands of refugees coming from Kosovo to Albania in March 1999, to escape the barbarisms of the Serbian army inspired by the regime of S. Milosevic who had destroyed their lives. My employment as project manager in February 1999 by the AMMF coincided with the commencement of Kosovo exodus one month later. On March 25th, people had come out of their homes and from their courtyards saw the dark sky and listened to the noisy whir of NATO aircraft passing for the first time over the city to bomb the Serbian military objectives. Glenn and Martha were visiting my home that night and from the yard of the house, looking up at the sky we heard this "spectacle" of whirs. Five days later, on March 30th, Glenn called me for a meeting to draft a concrete aid plan for Kosovo refugees who were arriving very fast in Lushnja, as in all of the Albania. Their number in Lushnja district reached 6,000, and were sheltered in Lushnja families, in the dormitory of the mechanical school, and in some improvised large camps, such as in the former Car and Tractor Station (CTS) building in Lushnje, in Divjaka beach booths and in Kolonjë.

The Mennonite Foundation assistance focused on several aspects: providing two field kitchens with capacity of several hundred people, arriving at Rinas Airport from Switzerland, providing mattresses and blankets, providing a large amount of canned meat sent by the MCC and EMM in America, which arrived in Greece through Germany, but in Thessaloniki the amount of meat was blocked for over two weeks with the claim of lacking documents. The meat was fresh and every box had the expiry date. After multiple calls and clarification of the situation by Glenn, finally everything was resolved and two big trucks of a Greek company loaded with canned meat passed the Albanian border and arrived at their destination in Lezha, to be distributed as humanitarian aid to Kosovo families. Another problem came up by the truck drivers who required each $ 1,000 for transporting the meat, a problem solved by Glenn and Paul Kropf, who explained to the drivers that the company owning the transport trucks had been paid in advance for all the services, including transportation, and they would be paid by their company.

Large quantities of clothing, brand new home appliances, toys for children, etc., came from USA, Germany, England, the Netherlands and Switzerland. Thousands of packages from these countries which were collected in Germany, were then sent to Tirana and from there were distributed to various cities of Albania through humanitarian foundations and churches; a significant amount also arrived in Lushnje, where a well organized distribution took place in cooperation with the District Council and the Municipality. In many of these packages, especially those with items of personal use, toys and school articles, were also letters addressed to children and Kosovo families, written by the American citizens who had prepared and donated these packages, showing a meaningful example of solidarity, friendship and encouragement in those extremely difficult days for the refugees from Kosovo. For the distribution of these aids, the Mennonite Foundation rented premises to carefully store all the aid boxes and coordinated the work with the Municipality of Lushnja and other organizations and churches that worked for the same purpose, in order to prevent duplications in different families and areas where refugees from Kosovo were sheltered. The city was divided into several areas belonging to different humanitarian organizations. The Mennonite Foundation created a special center for aid at a rented warehouse called the "HOPE Center". In addition to the maximum engagement of the American missionaries working in Lushnje, 7 people were hired (of whom 4 were women working in the aid warehouse), as well as a van driver for the transportation of goods. Dan and Melike Smenge designed and printed a special coupon for receiving the aid, such as clothes, food, toys, or household appliances. This coupon with the stamp of the AMMF was given to every Kosovo family in the respective area with the right to present it to the aid warehouse and to receive themselves the goods they needed. The coupon contained 30 squares and each had the value of a point (*photo no. 11*).

By marking a square, the holder of the coupon had the right to choose an item, either clothing, shoes, or toys for children or household appliances. Foods (oil, flour, pasta, sugar, beans, rice, cans, etc.) were delivered directly by the foundation employees with their cars to the houses or places where the Kosovo families were accommodated. This was done on the basis of a predetermined weekly calendar, voluntarily. In this way, close contacts and conversations with Kosovo families were established. Often they showed the painful stories of their sufferings. One of the buildings where the Kosovo families were accommodated was the dormitory of mechanical school. In this dormitory were sheltered 110 people in total, all of whom were furnished by the Mennonite Foundation with 150 mattresses, blankets, food, personal hygiene items, sanitary materials for children and medicines.

In one of the meetings held on 3 April 1999 with the main local authorities, the Prefect of Fier P. Arapi, the Chairman of Lushnja District Council Landi Bano, the Mayor Agim Fuga and the Head of Police Station V. Çela, the representatives of humanitarian organizations present in Lushnje, Caritas, Mennonite Foundation, Red Cross, Evangelical Community, Cafod etc. received information on the real situation of the refugees, the needs of the families from Kosovo, the emergency measures to be taken and the coordination of works. Until that day, 204,000 people had arrived in Albania, of whom approximately 4,000 in Lushnja district. The Chairman of the Council, Mr. Landi Bano assured us that for the supplies and the deliveries the State would provide vehicles, while the police would escort them if necessary. The urgent needs of the camp established in Kolonjë were for 1,000 mattresses, 2,000 blankets, while for kitchen utensils we were given an accurate list of plates, spoons, forks, pots, cups, etc., 3 pumps for the provision of drinking water, as well as the construction of toilets. Caritas was ready to help out there immediately. In the next meeting three days later, we learned that the number of registered missing people from Kosovo in Lushnja district reached 5484 people. A major problem for the newcomers from Kosovo was that many families were separated and they did not know where to find and how to communicate with the other family members, brothers, sisters, or cousins. They gave us names, phone numbers or addresses. For example, S. Gashi from Krusha e Madhe was in "Kadri Qystri" neighborhood in Lushnje with his wife L. Gashi; his brother's family was in Tirana and he was unable to communicate with them, while his two sons, three sons of his brother and a grandson had been arrested by the Serbian police and he knew nothing about their fate, if they were alive or killed.

The entire Evangelical Community in district of Lushnja made a great contribution, staying close to the refugees from Kosovo, knowing and completing the needs for humanitarian aid. Berti Dosti was one of the leaders of this community, who worked tirelessly to distribute these aids directly to many Kosovar families.

Solidarity with the people of Kosovo those days exceeded Albania's borders. In my notes of those days I have written that the Italian show of Bruno Vespa "Porta a Porta" on the evening of 13 April 1999 was titled "Insieme per il Kosovo" (Together for Kosovo), in which several Kosovo refugees were invited. The 53rd Prime Minister of Italy Massimo D'Alema was interviewed, who strongly denounced the Milosevic's regime and the barbarisms against the innocent population in Kosovo, while a group of RAI personalities including Carlo Conti and Fabrizio Frizzi, worked for an open account for the Arcobaleno Mission with phone calls to assist the Kosovo families in Albania; the amount grew very rapidly and reached 4 billion lire. During the show, on a big screen were displayed sad pictures and videos of exhausted Kosovo people, under the soundtrack of the Slave Choir from Opera “Nabucco” of Giuseppe Verdi.

Then the interpretation of the song “Liberta” by the famous Italian singer Albano Carrisi created meaningful emotional moments.

5. THE FARMERS' FEDERATION OF MYZEQE - FFM

5.1 Challenges for the establishment of Farmers' Federation

Let's go back to the registration history of the Farmers Federation with central office in Këmishtaj, which we mentioned earlier. The work for its establishment started a year earlier, in 2003, when the AMMF began meetings with the governing boards of three associations, F1 in Goriçaj, F2 in Mërtish and F3 in Fierseman. At the meetings with the F2 and F3 associations, since the beginning, their governing boards showed willingness, even enthusiasm for cooperation. While with the F1 association the work proceeded slowly. Being the first established association and with a busy activity in the field of community development, during multiple meetings, its leadership had the tendency to show the experience and superiority over the other associations and wanted to be favored in the future federation. In fact, the successful history of the first complex project in Goriçaj is a joint tireless work, of both organizations, the Mennonite Foundation and the F1 association. Commitment and cooperation between them has been to a high level. Regarding the second project, involving more than one of the villages of Gradishtë Commune, article 8 of the ICCO Draft Document Summary (a document which was translated and distributed to the associations along with other documents) clearly defined the position of the participating associations in the project: *"It is crucial for the success of the project that all three associations participate equally in the planned activities. None of the associations has a favored position".*[4]

It seems that this point was not well understood by the leadership of the F1 association, which in fact had the deserved place in the new project, in all its activities. The first president of the Farmers Federation of Myzeqe was elected one of the leaders of the F1 association, Trifon Stambolliu, a capable, hard-working farmer and very active in the new developments in the agricultural sector. He held the position of the FFM president, according to the statute, for the first 3 years.

The first joint meeting with the three governing boards and the representatives of the Mennonite Foundation headed by Daniel Hess, took place on October 5, 2003 in Goriçaj village. In this meeting were presented the main objectives, components and activities of the new project entitled "Increasing the income of farmer families through strengthening the capacities of farmer associations and improving their organizational structure" co-funded by ICCO in the Netherlands. The document was drafted by Professor Astrit Balliu. ICCO contribution to this project was € 136,360. During the first meeting was presented the model of the cooperation agreement between the Mennonite Foundation with each of the three associations, which contained the main points of cooperation within the new project and the obligations of each party *(Annex 6)*. ICCO had defined as a partner organization and project contract holder, the Mennonite Foundation (AMMF). After long discussions, the agreement was supported by two associations, F2 and F3, and was opposed by F1

---

[4] ICCO Project Document Summary, Annex 3, article 8, year 2003

president, who expressed reservations about cooperation with AMMF and requested more time and details for the project before giving his consent to the agreement. This objection, although not expected, was taken into consideration by the AMMF, which agreed to complement the information about the project with more detailed data, especially regarding specific activities. Also, the main co-financing agreement between ICCO and the Mennonite Foundation as a contract holder, was translated from English into Albanian to be given to each of the three boards of associations at separate meetings with them. So, after more than two weeks, on October 22, 2003, two other meetings were held, respectively with the governing boards of the F2 association in Mërtish and the F3 association in Fierseman, to whom were presented other detailed documents related to the project. Both boards, led by their presidents Stavri Gjini and Petraq Çapuni, continued to show full readiness to work together and signed a cooperation agreement with the AMMF, which was represented at the meeting by Willard Heatwole and A. Rrapokushi.

The meeting with the board of the F1 association took place two weeks later, on 3 November and, just as with the other two associations, besides what was said at the first meeting, were presented other materials related to the project with detailed information on its seven main directions:

1. Creation and opening of the management offices of 3 associations.
2. Creation of Market Information Network.
3. Increasing the trade skills and opportunities of associations.

4. Strengthening the Association's Agricultural Extension Service.
5. Capacity building of associations board members and staff.
6. Supporting the establishment of Farmers Federation.
7. Project management.

However, the agreement with this association, despite the opportunity offered to be included in such an important project, was not signed that day, because this time its leader (R.Th.) requested the signing of the agreement to be postponed for another date, as he wanted to know in advance the fund that the F1 association would benefit from the project for each activity. Such detailed information, with the project just started was not easy at all, given that funds for these activities belonged also to two other associations. Despite these difficulties, the Mennonite Foundation was willing again to cooperate and at the next meeting held on November 12, 2003, both sides agreed and signed the agreement between the F1 and the AMMF, leading to full implementation of the project. One of the objectives of the Mennonite Foundation was to involve the farmer associations in implementing the project to a wider and more active scale, so the signing of the agreement with them was very important. Let's briefly stop on a specific problem related to the leadership of farmer associations: the president of one of the associations shows up in two different scenarios in two different projects (although from the same donor). Firstly, in the ICCO funded big rural community development project, he is a skilled leader, who manages competently the association's governing board, engaged at a high level in his duty and cooperating with key stakeholders, including the Mennonite Foundation, from the beginning to the end of the project, either in the case of the construction of the village water supply or in the case of road rehabilitation as well as in smaller projects of agricultural profile. The cooperation between these two organizations, F1 and AMMF, and the important role of the latter, is also highlighted in an ICCO assessment mission conducted from 5 to 25 July 2002 by the professors of the Agricultural University of Tirana, Fatbardh Sallaku and Ilir Kristo, that states: "*The implementation by F1 of all the projects supported by ICCO, has been assisted and inspected by the AMMF (Mennonite Foundation). The presence of a qualified AMMF staff and the daily assistance provided by this staff has enabled the successful implementation of all the ICCO-supported projects*".[5]

While in the second project, for the development of a much wider community involving not only one but at least 6 villages of Gradishtë commune (Këmishtaj, Goriçaj, Mërtish, Fierseman, Gradishtë and Babunjë), there is lack of readiness for cooperation from one of 3 associations, since the beginning of the project there is some reluctance from its leadership, causing the process to slowdown. The Mennonite Foundation, which was a promoter and strong supporter of the association in question and the warrantor for the water supply and road rehabilitation project, expected another, more active and cooperative approach. In fact it was the opposite, the leader's position affected the majority of the association's members throughout the project, where the extent of its involvement after the signing of the agreement was not at the expected level, but at a lower level, despite the concrete

[5] "Community Development Program Evaluation Mission" F. Sallaku, I. Kristo, p. 3

efforts to bring it closer and to include it as widely as possible. In view of two facts, the first that the F1 was the largest association in number and the second, that it had the largest surface of greenhouses, over 10 ha in 2003, the number of farmers from this association who supported the Farmers Federation of Myzeqe in the marketing of agricultural products, bringing their products to its warehouse was small. But in addition to the above facts, there is another important factor, that cannot be neglected and should be considered, the market: a big part of the production of Goriçaj greenhouses was sold in this village, traders from different towns of Albania came directly to the farmers' greenhouses with their cars and made individual oral arrangements to buy qualitative products, while some other farmers sold their products to the market of Lushnja. The AMMF took into account the market factor and respected the arguments, but if the leadership of the association would have shown a greater commitment in the framework of the new project, this would had significantly affected the increase of sales, as a wider and well-organized group, the improvement of sales standards and the enhancement of competitiveness. The best participation of the F1 association in the joint marketing was in 2007, when up to 23 farmers from Goriçaj (not permanent throughout the year) were included, bringing 40,183 kg of production, which was only 5,5% of the joint sales within the Federation. Commenting on the fact of the involvement of a small number of farmers from Goriçaj in 2005 in the joint marketing led by the Farmers Federation, the president of the F1 association at a meeting held on 5 July of that year with the participation of Dutch consultant Gerrit Holtland, honestly stated: *"It's a little hard to talk, because we as an association have not brought our production in the warehouse. This is because traders from all over the country, from many districts, come directly to our village. The farmer's mentality is still under the example of work, that we, as leaders have done".*

5.2 Promotion of the idea of Federation and making relevant decisions

During November 2003, the farmer associations through their governing boards worked to promote the idea of establishing the Federation. It was followed by other concrete steps by organizing several meetings of the General Assembly for a really good start of the project. In these meetings, every association decided to establish a Farmers' Federation named "Albanian Farmers Organizations Union". The agenda included two points:

- Participation in the establishment of the Albanian Farmers Organizations Union.
- Full rights membership in the Albanian Farmers Organizations Union.

Meetings of the assemblies of the three associations took the relevant decisions for accession to the Farmers Federation. Within November, respectively on 13, 20 and 24, these decisions were notarized by the F1, F3 and F2 associations respectively to submit to the Court of Tirana. Such a decision was made by the Board of Directors of the Mennonite Foundation, but which actually was not submitted to the Court of Tirana, upon the advice of the Dutch project consultant Gerrit Holtland, who rightly suggested that the Mennonite Foundation, as an organization should stay out of the farmer's union. Gerrit also made a reasonable remark about the name that was chosen:

Albanian Farmers Organizations Union - a bombastic name, not suitable and realistic - replacing it with the simple but acceptable name by all the actors in involved the process: Farmers Federation of Myzeqe (FFM). On the other hand, the Mennonite Foundation was the signatory of the contract with the donor and being the project manager, with the assistance of the accounting expert, Mrs. Mimoza Sula Aliaj in Tirana, drafted some important documents, such as the regulation of operation of farmer associations management offices, the regulation of the project headquarters in Lushnje, the document for the process of selection and training of the managers of each association, as well as the manner of administering the approved budget, including the mode of funding, the control of funds, the way of reporting and the way of budget implementation, which lasted 27 months from October 2003 to December 2005. Mimoza also provided a professional assistance in training the new staff of the three associations and of the Farmers Federation, by introducing the Albanian legislation in the field of NGO financing and well-management of project funds, a contribution that was very valuable at the initial phase of the project. Below is the text of the regulation on the manner of funds administration for farmer associations:

THE REGULATION

On the manner of administering the approved budget for the implementation of the project "Farmers' Federation and Marketing Program '03 - '05" for Farmer associations F1, F2 and F3.

1. The mode of funding

Each association will be funded for the amount envisaged in the budget plan by the AMMF Lushnje in quotas according to its budget and structure.

The funds will be in the currency specified in the co-financing agreement with ICCO and precisely in EURO.

The amounts financed by the AMMF will be transferred to the respective account opened at the Savings Bank Lushnje branch for each association.

2. The way of budget implementation

The use of budget funds will be in accordance with its structure.

It is not permitted to use unrealized funds from one part to another. In cases where the association has not realized some parts of its budget structure and thinks about changing the budget plan, it should request it in writing to AMMF. After reviewing the budget indicators and receiving the donor's approval for these changes, the AMMF notifies the association on the approved changes and after that approval the association has the right to use these changed parts.

The associations are responsible for making decisions about the project implementation in accordance with its activities.

3. The way of reporting

At the end of each month the associations are obliged to report on the implementation of the budget according to the defined format. The reporting currency should be the one in which the project is financed, converting it at the fixed rate of 1 EURO = 135.86 lek. The reporting deadline is within date 5 (five) of the following month.

4. Other income

Besides the approved budget, associations deposit funds from the contributions of the project beneficiaries to the bank account of the association in ALL. The use of these funds will be done at a second stage in accordance with the amounts of contributions paid out by the beneficiaries.

The Association may earn income from the exchange rate differences or bank interests. In case of loss from exchange rate under article 3, the association requests reimbursement of the difference to the headquarters. Currency exchange must be certified by a legal document.

Approved

Project Headquarters
Lushnje, 17 November 2003

5.3 Various meetings

The last quarter of 2003 was very intense, filled with meetings and activities not only in Lushnje but also in Tirana. On October 23, was held a meeting with the director of the Agricultural Service Project (ASP) at his office in Tirana, at the presence of Zef Gjeta, the well-known marketing specialist, Dritan Kolleshi, the director of the new wholesale fruit-vegetable market in Lushnja, Willard Heatwole, the rural development adviser from the Mennonite Foundation, and me. During the meeting, the ASP Director made a brief outline of this important project and suggested that farmer associations should submit a grant proposal for the selection and packaging of their agricultural products. He also said that the representatives of the FFM (Farmers Federation of Myzeqe) may also be members of the ASP board. However, these were not realized, despite the fact that the FFM submitted a complete grant proposal with all the documentation needed for a tomato selection equipment and certainly did not even become an ASP board member. By coincidence, exactly one year later, on October 23, 2004, in Këmishtaj came for a meeting with the farmers of Myzeqe area, the Minister of Agriculture of that time, Mr. Agron Duka, accompanied by Mrs. Arta Dade, deputy of this area in Albanian Parliament. At this meeting, which took place in the building of the former cultural center, I raised two problems: firstly, the reduction of the fuel price for the farmers of Myzeqe and secondly, the problem of application we had made as a federation to receive a grant for the tomato selection equipment. At the end, Minister Duka answered to all the questions and expressed his full support for the FFM Këmishtaj and its grant application by personally saying to the Grant Director Mr. Agim Sh., present at the meeting: *"The Farmers Federation should absolutely receive the grant. These are three farmer associations and the organization of farmers in Albania is the Achilles heel. Therefore, they should receive the grant ... You should also invite members of this federation into trainings ... ".* After the meeting, the Grant Director, Mr. Agim Sh., while we were walking together, grabbing my arm said, "Despite what the Minister said, you do not win the grant (!), because in the application as a service provider you introduced Tani (D. Kolleshi, the first director of wholesale fruits and vegetable market in Lushnja), which is not allowed and you have some mistakes in the project ...But I will invite you in my office in Tirana and we will talk" he said. It was a pity that such a grouping of farmers organized in an association was not stimulated by the State. The tomato selection machine, although not through the Albanian ASP grant, came to Këmishtaj two years later from Valencia, thanks to the Spanish project, with the help of Prof. Civici and Prof. Collado. The Spanish were more generous than the Albanians *(photo no. 12, no. 13).*

Three years later, on October 26, 2007, the other Minister of Agriculture, Mr. Jemin Gjana, in response to our invitation, visited the Farmers Federation in Këmishtaj, accompanied by a large number of his staff. Present at the meeting were the Director of the Spanish Project, Jose Luis Collado, and the president of Mennonite Foundation, Paul Kropf. On this occasion, the Federation had displayed the organic products of the pilot group, tomatoes, cabbages, cauliflowers, Chinese cabbages, peppers, etc. There was also a positive demonstration of tomato selection with the equipment that came from Spain. During this visit, Minister Gjana was attracted by the designs of the new warehouse project exposed to the wall, a project he favored and ordered the director of the Fier Agricultural Directorate, F. Troka, to see the possibility of giving our Federation a land surface for building this warehouse, in response to our request. However, as in the aforementioned case of the grant for the selection equipment, neither in the second case we were offered any option for land surface by Fier Directorate, despite the respective orders from both ministers (the first belonged to Socialist Party and the second one to Democratic Party) to support us.

5.4 Opening of Farmer Associations Offices

In order to strengthen the organizational and management capacities of the farmer associations, on November 22, 2003, at the AMMF office in Lushnje was held the procedure of interviewing and selecting the managers of the three associations. The commission, consisting of representatives from the Mennonite Foundation and three associations, selected 3 candidates, respectively:

- Ll. Stavri for "F1" association office in Goriçaj
- F. Ndoni for "F2" association office in Mërtish
- R. Rrushi for "F3" association office in Fierseman

Meanwhile, the locations of the offices for each association were determined according to their headquarters. The leaders of Gradishtë commune at that time, Vasil Sema and Ylli Stambolliu, provided a very good office for the F2 association on the second floor of the former cultural center in Këmishtaj, while the F1 association office was placed since the previous project in a special room at Bar Cafe Mondial of Trifon Stambolliu and the F3 association had begun building a simple office in Fierseman with the contribution of the president, P. Çapuni and of the project.

Each association opened a special bank account for ICCO project. Thus the financial actions between the project headquarters managed by the Mennonite Foundation and the branch offices were conducted through the bank. The allocation of funds was done according to the approved budget structure and the respective association was notified in writing by the project headquarters about the transaction conducted in the bank. As example, we are showing a notice of transfer of funds to the F3 account on December 3, 2003:

THE ALBANIAN MENNONITE MISSION FOUNDATION

No. 002 prot.

Lushnje, 03.12.2003

Project: "FARMERS' FEDERATION AND MARKETING '03-'05"

HEAD OFFICE

To: Office of F3 Association, FIERSEMAN

NOTICE

About budget structure.

Activity: Creation of associations' offices

Period: 1 October 2003 - 31 December 2003.

We inform you that today, on December 3, 2003, the project headquarters transferred to the account of the F3 association Fierseman, a fund of EUR 1490 (one thousand four hundred and ninety) for the opening of your association office in Fierseman.

The budget structure for using the above sum is as follows:

No.	DESCRIPTION	EUR	ALL
1	*Computer (1 piece)*	*700*	*95,102*
2	*DeskJet printer (1 piece)*	*150*	*20,379*
3	*Office equipment*	*200*	*27,172*
4	*Stationary*	*50*	*6,793*
5	*Office rent*	*90*	*12,227*

6	*Office manager salary*	*300*	*40,758*
TOTAL		**1,490**	**202,431**

Project Manager

(signature, stamp)

In December, started the training of office managers in the field of financial management. The managers benefited a monthly salary from the project. The salary indicated in the table, EUR 300, includes the quarterly period from October to December 2003.

In parallel with these actions, were taken other necessary steps to legalize the Farmers Federation of Myzeqe in the Court of Tirana. All the notarized documents of the three associations at the end of November 2003 were presented at the lawyer's office, Alfred Mustafa in Tirana, who finalized the statute and the founding act of the FFM. He started collaborating with the Mennonite Foundation at the beginning of the year and started drafting these basic documents in March, doing a professional and serious job and submitting in April the first draft of the statute after the suggestions he had received directly in two meetings in Lushnje from the four organizations involved in the process.

5.5 Delays for registration of FFM Këmishtaj

Creation of a broad-based representation NGO, where members would be legal persons (in the case of Federation, farmer associations from various rural communities of Myzeqe), was something new in 2003-2004 and this process of course had its own difficulties, related to the lack of such organizations in Albania, the farmer's mentality for such relatively large groupings, their leaders, and so on. Therefore, taking into account these factors, foreign experience and legal professional assistance was taken into consideration.

After the signing of the three cooperation agreements between the AMMF and the 3 associations on October 22 and November 12, 2003, it was time to register the Federation in court. But a month later, on December 12, 2003, another unforeseen obstacle emerged. On that day, three representatives of the Mennonite Foundation (among them the director, Daniel Hess), and the presidents of the three farmer associations of Goriçaj, Mërtish and Fierseman travelled to Tirana to sign the statute and the act of establishment of the Farmers' Federation of Myzeqe (FFM) before the notary, as well as to purchase new computers and printers for the offices of the three associations. While the second assignment was fully realized, all the equipment was purchased as planned, the first one, the signing of the FFM's basic documents failed to realize, because at our lawyer's office Alfred Mustafa in Tirana, although the final version of the statute and the founding act reflected the previously expressed opinions by the three associations at a meeting held in Lushnje, the leader of one of these associations (F1) disagreed with the names of people who were appointed to the Administrative Council of the Federation, so the notarization was postponed. All of us were disappointed from the attitude of this leader.

This process was completed only 6 months later, on June 5, 2004 after reviewing once again the articles of the statute, the founding act and the people nominated in the Administrative Council. The notary act was done in Lushnje and the decision of Tirana Court to register the FFM was taken on 26 July 2004. (*With Fred, the lawyer who was engaged for a relatively long period of time to draft the Federation's basic documents and to obtain opinions in the field, by chance after 12 years, in 2016, I worked in the same building, but in different floors, in the best area of former Block in Tirana. At that time I was working in the position of the administrator of consulting "PLANET" Company*). Other well-known experts, Gerrit Holtland, Willard Heatwole and others also made a valuable contribution to the establishment of the Federation.

But not always meetings with association leaders and board members were easy. This is due to the lack of experience in this area, because for almost 50 years, until 1991, there were no associations or genuine agricultural cooperatives, except the communist ones, which also affected the mentality of Albanian farmers. But I want to mention a fact that has to do with another situation created not only by the lack of knowledge, but also by the lack of professional ethics during communication of one of the leaders of the associations with the well-known expert Gerrit Holtland at the meeting of 12 May, 2004 in Këmishtaj; Gerrit described very well this situation in the email he sent after the meeting to professor Astrit, me and Daniel Hess, underlining: *"The meeting was dominated by the chairman of the association, who has no idea of what to do and on how to do it. He only thinks in terms of "who is fixing the prices". He said he wants to unite all farmers in Lushnja "to agree on the prices". He simply does not understand that in a situation like in Albania, nobody can "fix" prices (not to mention the fact that he never managed to get the members of his association to agree on one price). On top of this, it proven again that he is not able to listen; he did not listen to me...nor did he listen to anybody else in the room. For more than one and a half hour he spoke more than 80% of the time and I could not even managed to ask others about their opinion...What you (Andon, Astrit, Dan) want to do with this, is up to you. For me it is OK if you tell this to him or other people; it is also OK if you do not wish to do this...I have no bad or negative personal feelings against him; I believe he has a good heart. But based on my experience, in and outside Albania, I am only deeply convinced that his attitude is very counterproductive in any meeting and ultimately also for the farmers of the association*"[6].

At its beginning the FFM managed to include 227 members, small farmers, who belonged to the 3 farmer associations.

The Mennonite Foundation with its President Paul Kropf was the promoter of farmer associations cooperation in Lushnja region. Paul, though living in Lezha with June, his wife, about 100 km far from Lushnja, always supported rural development projects in the area of Lushnja, which he often visited. Having a clear vision for the rural socio-economic development and in alignment with its humanitarian purpose, the Mennonite Foundation with

---

[6] Gerrit Holtland, email to Astrit Balliu, Daniel Hess, Andon Rrapokushi, 12 May 2004

dedicated employees has done a tremendous job for years in the region of Lushnja, assisting dozens of farmers with micro-loans, but also with small projects and grants, helping these poor families to hope and to return working on their land by building a large number of solar greenhouses and creating small livestock farms with dairy cows or sheep herds. This new reality created a positive impact on the young people of Goriçaj and the surrounding villages, who now saw their future not only in emigration, in Greece and Italy, but also near their families. In addition to the field of agriculture, the humanitarian aspect of this foundation also contains in its archive the great humanitarian aid to Kosovo refugees as mentioned, which comprises a separate chapter in its history, as well as the great work that has been done in Lezha over those years thanks to the dedication of other employees, Allen Umble, Norma Telesh, Leon Zimmerman, etc. Let's turn to the project funded by ICCO, related to the implementation of the main activities of this project, Professor Astrit Balliu, also the drafter of the document, gave a valuable contribution in this regard. His work as an external consultant included the service with agricultural extension coupons, training on association business plans and marketing issues of agricultural products. He had been close to farmers, members of three associations and their leaders. But besides Professor Astrit, in the framework of training the associations leaders and staff were invited and contributed also the well-known specialists of the above issues and issues of cooperation of small producers, such as professor Maksim Meço, Shkëlzen Marku, etc. A special training program with 10 topics defined for business plans, marketing, organization of farmer cooperatives etc., which lasted for five months with two topics per month, from 20 March to 3 September 2004, was held in the meetings hall at the former cultural center in Këmishtaj with the financial support of ICCO - Netherlands and the organizing and managing assistance of the Mennonite Foundation, with Executive Director, Daniel Hess. The first training with the participation of three associations and AMMF was addressed to the way and cycle of project management, elaborated by Professor Shkelzen Marku. Such qualitative training, with genuine experts, was a qualification and education school for the farmer leaders of the area of Këmishtaj. They continued under the Spanish project, during which a large number of training courses for ecological agriculture were performed by the Albanian and Spanish experts, which will be mentioned later, as well as trainings on farmer associations management such as "Accounting for a farmer association" elaborated by Prof. Ass. Flutura Kalemi, some trainings on conflicts, developed by Professor Tomi Treska, etc.

Conclusion:

- *The establishment of the Federation was a great success of the farmers of Myzeqe area and of the three associations, which was achieved after many efforts, meetings with farmers and their leaders, and consultations with local and foreign experts. Establishing and registering the FFM was a process that the Mennonite Foundation managed and monitored with discretion.*

6. THE FFM FUNCTIONING AND JOINT MARKETING

6.1 Decision-making and governing bodies

The Federation from the legal point of view functioned as a non-profit organization but with the features of a regional marketing cooperative and had these governing bodies:

- Assembly of Members (consisting of 9 persons, representatives of three member associations, three for each association) was the highest decision-making body of the FFM association.

- Administrative Council (consisting of 3 persons, one for each association). This council contained three important positions: the president of the Federation, the vice president and a member. The Administrative Council had a very important role, it was responsible for the realization and administration of the Federation's activities, and in this context, it was entitled to make decisions for the smooth running of its daily activities. By secret ballot, since the second joint meeting of the representatives of 4 organizations (three farmer associations and the Mennonite Foundation), held on December 20, 2003 in Këmishtaj, resulted:

President of the Federation was elected Trifon Stambolli from F1 association with 3 votes in favor and one against; vice president was elected Stavri Gjini from F2 association with 3 votes in favor and one against and member was elected Andi Semini from F3 association with 4 votes in favor. This Council was reaffirmed at the General Assembly meeting that took place in August 2004 after the FFM's registration in Tirana Court. Other governing bodies were:

- Director, who was the legal representative of the Federation and headed its staff.

- Financial committee, a controlling body, composed of 4 persons, one from each member association and one external expert. In the composition of this commission, Petrit Toska, who was representative of the F3 Association and one of its leaders, made a professional work.

Federation's permanent staff consisted of the director, sales manager, financier, driver, and an agronomist, who was employed later.

Every decision taken by the Members Assembly of the Federation was signed by the presidents of farmer associations or a person authorized by them. Other challenges were ahead, and one of the most important, the marketing. The marketing component was the main pillar of ICCO project, therefore the biggest efforts focused precisely on dealing with it. In fact, during the process of establishing the FFM, before its registration, more precisely in March 2004, efforts were made to find a warehouse that would serve as a collection and sales point for the products of 3 associations members. This was enabled by renting a facility in Këmishtaj owned by A.M. The warehouse had a surface area of 310 m^2 and served for storing grain at the time of cooperative, before 1990. It was a very thick walled building, suitable for the intended purpose, but without ceiling and a lot of need for repairs. Both parties agreed on the amount of rent, and the funds for its repair were considered as part of the rent. Necessary repairs were made inside and outside the building based on the estimation presented by the specialist K. Rama, who also followed the works, transforming it completely. For these actions, the assembly of members of FFM made the relevant decisions on September 7, 2004. When the marketing activity started, the rent value was 40,000 ALL/month for a period of 31 months which was later renewed. This warehouse was not just a collection and sales point, it had some important functions during all the time of use, for more than four years. These functions were:

- Collection and sales point of agricultural products.

- FMM institutional center, where director and staff office was located.

- Information center for farmers of the area, where they received regular information on the daily prices of products, markets and agricultural inputs, new varieties, etc.

- Education and training center for farmers and their leaders, a special large room, adjacent to the staff office.

6.2. Poll organization

In order to get the views of the farmers of the area about collecting the products at a joint collection point and, most importantly, choosing where this point should be located, as well as other aspects of marketing, such as packaging, selection, etc., a survey of 9 questions was distributed to the members of the three associations in June 2004. A total of 73 farmers from different villages participated in this survey. The organization and collection of results was done by association managers. The questionnaire contained the following questions:

1. Do you agree to have a joint collection and sales point of fruit and vegetables for the three associations part of the Farmers Federation?

Yes No

2. Do you think Këmishtaj is the most location to accomplish this marketing project?

Yes No

3. Do you think this project will increase opportunities for a better marketing of your agricultural production?

Yes No

4. Would you agree to use a new packaging method, with boxes that respond to market demands today and tomorrow?

Yes No

5. Do you think that collecting and selling on behalf of your association, will increase the opportunities for new markets in Albania and abroad?

Yes No

6. Would you be willing to contribute to this marketing project?

Yes No

7. Do you think traders will have interest to buy our products collected in one place?

☐ Yes No ☐

8. Do you think that this new way of collecting and trading will ease the transport problem and save your time spent in Lushnja market?

☐ Yes No ☐

9. do you think it might be better to s production, at the warehouse or where it is produced?

____

This form is distributed by the farmer association (name of association ______________________________).

Farmer's names were not required.

The results of this survey are shown in the table below:

QUESTION	ANSWER OF "F1" ASS.		ANSWER OF "F2" ASS.		ANSWER OF "F3" ASS.		ANSWERS IN TOTAL	
	YES	NO	YES	NO	YES	NO	YES	NO
No. 1	30	1	18	1	23	0	71	2
No. 2	17	14	18	1	23	0	58	15
No. 3	30	1	18	1	23	0	71	2
No. 4	30	1	18	1	23	0	71	2
No. 5	30	1	18	1	23	0	71	2
No. 6	28	3	13	6	16	7	57	16

No. 7	30	1	18	1	23	0	71	2
No. 8	30	1	18	1	23	0	71	2
No. 9	7 in wareh ouse	24 in greenh ouse	2 in wareh ouse	17 in greenh ouse	6 in wareh ouse	7 in fie ld	15 in wareh ouse	58 in greenh ouse or field

The results of the survey were presented and analyzed at the first meeting of the FFM Members Assembly that was held exactly one month after being registered at Tirana Court.

Conclusions derived from the analysis of answers provided by the survey:

- *First question: only 2 out of 73 farmers were against a joint collection point. All the three associations are in favor of a joint point.*

- *Question no.2, related to the location where this joint point should be, along with question no.6, where the personal contribution for this project is required, received the largest number of negative responses, respectively 15 and 16, out of 73 in total. But distribution of these answers is different.*

- *In the second question, out of 15 responses against the place (Këmishtaj) where should be located the joint marketing point, 14 belong to F1 association. The explanation for this lies in the fact that the majority of Goriçaj farmers wanted the collection and sale warehouse to be in their village, arguing this with the fact that they were the oldest association, the largest in membership and with a large number of solar greenhouses. But if we look at this issue from another point of view, that this marketing center would serve not just a village, but at least 6 villages of the commune (Goriçaj, Këmishtaj, Mërtish, Gradishtë, Fierseman and Babunjë) with high agricultural production, as it actually served, then it is clear that the choice should have been different.*

- *Out of 73 farmers, 58 of them or 79.5% agreed that the collection and sales point to be in Këmishtaj. In favor of this were also 17 farmers from Goriçaj. Këmishtaj was favored by its geographic position very suitable in this case, but also by the ability to find a warehouse.*

- *In the sixth question, there are 3 farmers from F1, 6 farmers from F2 and 7 farmers from F3 association who were not ready to support the joint marketing project. However, in percentage, they accounted for only 22% of the total number of respondents, while 78% supported this project.*

- *Other questions on packaging and increasing the opportunities for new domestic and export markets received a positive response to a high degree as well.*

6.3 Opening the collection and sales point and strengthening the commercial aspect of the Federation

Prior to opening the collection point, it is worth mentioning an offer we had from a Fier-based company to rent the warehouse we were rehabilitating as collection and sales point for the agricultural products. Its representatives insisted several times to rent this facility. Since we had been working for a few months to return the warehouse to an appropriate facility with all the necessary premises, we asked the company in question to pay to the Federation 1 ALL/kg for all the quantity of production sold by the warehouse. This request was not accepted. At the second meeting we asked the company to pay to Farmers Federation a lease of 60,000 ALL/month. Not only was their offer ridiculous, 15,000 ALL/month for a newly rehabilitated warehouse with a surface area of 310 m^2, but even if this offer was somewhat higher we would not risk the project funded by ICCO. The Mennonite Foundation team (D. Hess. S. Harnish, A. Rrapokushi) who participated in the talks in the role of the project manager, did not accept the offer, despite that 1 or 2 people from one of the associations were in favor of it.

The Annual Meetings of the FFM Members Assembly were held every August, starting from 2004. In these analyzes participated representatives of the 3 member associations, FFM staff and usually were invited also employees of Mennonite Foundation.

In the first meeting that took place in Këmishtaj on August 26, 2004, a decision was made for the official opening of the Federation's activity and the celebration of its establishment. It was decided to be on October 30, 2004, and a commission for organizing this celebration was appointed.

THE AGENDA OF THE FIRST MEETING
OF THE FFM MEMBERS ASSEMBLY

Këmishtaj, 26.08.2004, Thursday, 10:00 A.M.

1. Registration of Farmers Federation of Myzeqe in Tirana Court. Reconfirmation of Federation Administrative Council and Financial Commission.

2. Federation's fundamental documents. Drafting internal regulation.

3. Official recognition of Federation, celebration of its establishment. Defining the date of celebration, the mode of organization.

4. Warehouse project. The current status of the facility, the improvement of its conditions, the need for repair. Contract with the owner.

5. Introduce Assembly to the performance of the project in general (by the project manager and the association's managers).

6. Questions, discussions and suggestions

As emphasized above, as the official opening date was decided October 30th , and the commission set up to organize this event was composed of representatives of the 3 associations and the project manager. This commission drafted a special program which included on its agenda:

- Official opening of the FFM activity with farmers from the area and guests from various organizations, local government and Ministry of Agriculture.

- Distribution of certificates of the training program of leaders and board members of farmer associations.

- Photography exhibition titled "Marketing – The Challenge to Win!".

- Organization of a fair with agricultural products of members of the 3 farmer associations.

- Awarding three prices for the best products.

In the official opening ceremony of the FFM activity came to participate the Deputy Minister of Agriculture Mrs. Vjollca Ibro and the AMMF President Mr. Paul Kropf (*photo no. 14 and 15*).

In the welcome speech, I emphasized: "*The establishment of the FFM represents an achievement for the farmers of this area, but on the other hand, marketing poses a challenge for the FFM and its 3 member associations, which consist of about 220 member farmers. Each of the farm associations has available an approved business plan, and they have the duty to turn this plan into a daily action plan ... The Federation is offering its members more opportunities to enter a new road of agricultural products marketing, offering them a fresh vegetable collection point, using its logo in labeling and packaging in small cardboard boxes, for a unique packaging. This is the beginning of a new and not easy job ...* "

After casual greetings, Mrs. Vjollca Ibro and the director of the Federation cut off the ribbon of the official opening of marketing point in Këmishtaj, which was the first in Albania of its kind. The photography exhibition in the warehouse premises was filled with the activities of the 3 farmer associations, while the fair was held with the best agricultural products and several prizes were awarded for the best products. A few weeks later, on December 23, Mrs. Vjollca welcomed me to her office in the ministry, where we talked about the benefits of the Free Trade Agreement signed by Albania with neighboring countries, the marketing point in Këmishtaj and the grant proposal that the FFM had submitted to the ASP. Like Mr. Agron Duka, the Minister, she expressed support for this grant and said she had contacted the ASP to consider our proposal...

The FFM logo, which was presented on the inauguration day, was meaningful, it contained three leaves in a circular form with the inscription "Farmers Federation of Myzeqe". Cardboard boxes with this logo were produced at the cardboard factory in Durres and Korça, but the company which offered to the Federation small cardboard boxes from the very beginning was AGROKONI. This company was connected to the FFM and cooperated very closely with it throughout the project for more than 4 years, since the first opening day of the warehouse on October 30, 2004 until December 31, 2008. This company run with competence and professionalism by its president, Ruzhdi Koni, is one of the most successful in Albania in the field of marketing of Albanian agricultural products. Its success is guaranteed by following a clear marketing strategy with Albanian agricultural products and embracing and implementing new agricultural technologies in the field of fertilizers, pesticides, seeds and seedlings.

Two and a half months before the collection and sales point was opened in Këmishtaj, in the middle of July 2004, together with Willard Heatwole, we went to Tirana and talked to the Administrator of Agro Koni, Ruzhdi, to establish a stable relationship between his company and Farmers Federation for the sale of agricultural products. Ruzhdi was immediately willing to collaborate. Professor Astrit, who was present at the meeting, estimated that only 10% of the total production of our 3 organizations required 20,000 boxes for three months. Such a cardboard box contained 8 kg of tomatoes and the use of these boxes by the FFM with its logo was a novelty, as almost all farmers of Lushnja at that time, but even later, in the wholesale market of the city traded with large plastic boxes of 20-25 kg. The production capacities of the associations were very large, but it was estimated that not all the farmers would bring their products to the warehouse. Business plans were real, though they were not estimated at the right level and equally by farmer associations. In marketing, since the early days, stand out the F2 association, which strongly supported the joint collection and sales project. This is shown in the figures and reports of that period. How is this support explained? The answer is simple, this marketing project was strongly supported by the president of this association, who was an experienced farmer and owned 3000m^2 greenhouses and as a consequence a large amount of product. He influenced with his example a large number of farmers in Këmishtaj-Mërtish area, who created a consolidated group that for several years brought their products into warehouse as members of the F2 association. The main products were tomatoes and cucumbers of greenhouses, as well as other quality products: peppers, eggplants, cabbages, cauliflowers, pumpkins, etc. and fruit such as peaches, lemons, strawberries, plums, etc.

On 29 August 2005, the 2nd Annual Meeting of FFM Members Assembly took place. The tasks and objectives defined by the Assembly at this meeting for one year from September 2005 to August 2006 were of an organizational, promotional nature, aiming at strengthening the commercial aspect. Functioning of farmer associations and their confrontation with the new Albanian reality in agriculture was set out as a priority, planning a seminar for this purpose, which was carried out at the end of September 2005. This annual meeting also assigned concrete tasks, as:

- Increasing the number of farmers who brought their products into warehouse and the improvement of marketing elements.

- Creating new connections with potential traders.

- Strengthening and expanding work with the group of organic agriculture farmers.

- Realizing a TV program to promote the marketing activities of the FFM in October 2005.

For the expenses covered by the FFM, the Members Assembly set the following quotas:

1. Payment for fuel costs for work trips within Gradishtë Commune: 500 ALL.

2. Fuel costs for work trips in Lushnje: 700 ALL.

3. Fuel costs for work trips in Tirana: 2000 ALL.

4. Payment for the sales manager working on Sundays: 1.000 ALL.

5. Payment per working day of Federation Finance Commission: 1.000 ALL/person (when control is performed and the respective report is compiled).

The joint marketing of the 3 associations started on October 30, 2004 and during the time of the project the amount sold through Federation generally grew. One of the best years in marketing for Federation, with the most complete data, was 2007, when the quantity sold amounted to 730 tons of vegetables and fruit. But even during 2008 there has been an increase, as per data published in the book "The cooperation in agriculture, an opportunity for production growth", referred to below. The quantity sold over a year was certainly not uniformly distributed over the months, but significantly higher during the greenhouses season.

The FFM has often exported greenhouse products, tomato and cucumber to neighboring countries Kosovo, Macedonia, Montenegro and Serbia at a better price compared to the local market. For the first time on 25 October 2006, 15 tons of tomatoes were exported to Montenegro at a price of 5 ALL higher than Lushnja market. During November-December that year, the FFM exported more than 70 tons of tomatoes to Montenegro, which brought a rise in the price of tomatoes, resulting in an increased income of farmers and Federation. The FFM exported to Italy at the beginning of December 2007 a load of 20 tons of tomatoes, with high-quality boxes brought by the Italian customer itself, but which had a delay of almost a week on its withdrawal. This delay became very disturbing for the Federation staff, and we gathered every evening to resolve this problem with the Italian company as soon as possible. It was on one of these nights after the meeting, as I was returning home with the Federation's technician Semini, that another car hit us very badly at a great speed, costing me two operations in the right hand, one in Tirana and another in Athens. Being a considerable amount of tomatoes, with a relatively high price, the value in ALL was too high, and understandably the responsibility that the Federation had towards the farmers who were expecting to be paid.

Conclusions:

- *With the official opening of the FFM activity began the normal functioning of its decision-making and governing bodies. Members Assembly played its part in strengthening the commercial aspect of the FFM and set quotas for the expenses covered by the Federation.*
- *Establishing sustainable connections with Agro Koni as the FFM main customer for sales and buying of inputs was of strategic importance to the Federation.*
- *Vegetable exports were important for increasing the sales, improving the quality and increasing the prices.*
- What was the model that the FFM implemented in marketing?

For this, we will refer to the Gerrit Holtland's missions in Këmishtaj, the Federation Dutch foreign consultant who gave a great contribution for the progress of Farmers' Federation during the whole ICCO Project.

7. THE GERRIT HOLTLAND MISSIONS AND FEDERATION PRIORITIES

7.1 Defining FFM priorities

The first mission of Gerrit Holtland, our Dutch consultant in Lushnje, took place from 12 to 22 January 2004. He dedicated separated working days to meetings with each farmer association and the Mennonite Foundation representatives. Apart from speaking well Albanian, a distinctive feature of Gerrit was his excellent communication with farmers, whom he loved, asked a lot, talked and understood. Gerrit had very clear ideas and he set the milestones for the establishment and the functioning of the Farmers Federation of Myzeqe. He defined the Federation's main priorities, insisting on three trends which gave to FFM the features of a marketing and service co-operative:

1) providing services in the field of agricultural products marketing, including collection, storage, selection, packaging and sale;

2) supplying farmers with agricultural inputs and fertilizers;

3) continuous advisory service.

His missions at the FFM throughout the project have highlighted both the achievements and the weaknesses of the associations and the Federation activity and in his reports Gerrit has provided valuable recommendations for improving the work and achieving the desired results.

7.2 Marketing model and different approaches to it

The leaders of the FFM Këmishtaj members associations, from 6 different variants of vegetable fruit marketing (collection, packaging and sales) presented by Gerrit at the meeting on January 17, 2005, after discussions, excluding two variants resulting in loss, chose the one of keeping 3% commission of each farmer sales income and 100 ALL / 300 kg production, as a weighing fee for each farmer. However, this was a stimulating model for the Federation to generate revenue and cover a large part of its expenses, but also for farmers and associations, who benefited a certain percentage from the revenue distribution.

Shortly, the 6 variants presented by Gerrit are as follows:

Variant	Percentage of production sold through Federation			Commision applied by Federation	Daily amount sold (in kg)	Result (ton /100 working days)	Profit / Loss in ALL for 100 Days
	Shoqata F1	Shoqata F2	Shoqata F3				
1	10	25	20	4%	63,2	632	262.327
2	5	10	10	4%	30	300	-485.953
3	5	10	10	7%	30	300	-107
4	30	40	40	4%	148	1.480	2.168.179
5	30	40	40	2%	148	1.480	530.972
6	10	30	30	3%	73	730	68.265

From the table above it is noticed that the best variant was no. 4, according to which the Federation had to keep 4% commission and the 3 member associations had to bring respectively 30, 40 and 40 percent of their production into warehouse, which their leaders could not guarantee.

In another mission in the first week of July 2005, were highlighted the results achieved by the first mission of the marketing and advisory service and the differences between the three associations in terms of their involvement in the project. During the meetings with the associations and the FFM, Gerrit was accompanied by the representatives of FMN, Willy Keener and Sonya Harnish, who were very committed and interested in the success of the project as the Foundation was the contract holder with ICCO. The meeting with the F1 association board held on July 8, clearly showed that they were not interested in bringing the production to the FFM warehouse in Këmishtaj, but would continue to sell it directly from their village as they already did, as the traders went directly there. At this meeting they asked ICCO to fund the construction of a small market in their village (an idea they came up with, after the first ICCO project). The second meeting that took place with the F2 Association board showed that this association was supporting the project, and especially the marketing organized by the FFM. As result of the service offered by the Federation to collect and sell the products, the F2 membership grew rapidly. Its membership in 2005 reached 73, all of whom brought their products in the warehouse and were pleased with the service provided by the Federation because the sales were ensured, the sale price was good and they did not waste time at the market.

The advisory service for the F2 members also worked well, selling 21 coupons and creating an income of 105,000 ALL for their association. At the third meeting of the Gerrit mission in July with the F3 board in Fierseman, its president admitted that the joint marketing project was not widely supported, but promised that at least 50% of tomato production in the second culture would come to the warehouse. However, the importance of this association for products from greenhouses was much lower than the other two associations. Gerrit encouraged the increase of production coming into the rented warehouse. Less than a year had passed since the beginning of the joint activity in this facility under the umbrella of the FFM, just 8 months, but it was clear who was the interested in this activity and who was not.

From the analysis conducted in July 2005 it turned out that the average daily output of the warehouse for 3 months, from April to June was low, 820 kg, but the capacities were much higher. In total, during 2005, as the first full and inexperienced year of the joint marketing, *were sold around 291 tons of vegetables, while in the second year this amount reached 661 tons, in the third year 2007 it reached 730 toms, in 2008 to 816 tons and in 2009 were sold 750 tons*[7]. Taking into consideration that Federation was focused on greenhouse products, the sales were very low from January to March and from August to September, with the largest sales during greenhouse season. Thus, despite the fact that the average daily for the entire year 2007, for example, was 2110 kg, in fact during the peaks season of greenhouses were sold on average 7-8 tons of production per day and 15-20 tons per day during exportation days. It was a whole new experience for everyone.

Given the current situation in the warehouse, the real facts and figures show the interest and predominance of the F2 association in terms of products sold through the Federation and by the opinion of the leaders of the governing boards of the three

---

[7] "Cooperation in agriculture" f.139, F. Allmuça, L. Kadiasi, Sergio V. Mulas, Tiranë 2009

associations, in a report drafted in 2005, Gerrit stated: *"... I think the arguments presented to give the warehouse building to the F2 association are still very valuable.* [8]*"* This document was presented at the FFM Administrative Council meeting on 12 December 2005, but his opinion regarding the ownership of the warehouse was rejected by the leaders of the F1 association, saying it would leave the F1 association outside the project. And in fact, the new warehouse built 2-3 years later was recognized as property of the Federation and not of the F2 association, but on the other hand, the contribution of the F1 association to this Federation remained low, as shown on the table below. This can be explained by the fact mentioned earlier that for farmers of Goriçaj the market was not a problem as traders with their cars bought the greenhouse products directly in the village and their production was of good quality.

The table below shows the contribution of each FFM member association during 2007:

Association	Center of association	Amount in kg.	Percentage
F2	Mërtish	567.578	77.8%
ADRIATIK Remas	Remas	90.635	12.4%
F1	Goriçaj	40.183	5.5%
F3	Fierseman	20.583	2.8%
PROGRESI	Gorre	7.299	1.0%
SHPRESA	Këmishtaj	2.518	0.3%
MIRËQENIA	Bubullimë	1.526	0.2%
Farmers' Federation (total)	**Këmishtaj**	**730.322**	**100%**

[8] "Some considerations on the future of FFM" Gerrit Holtland, December 2005

Certainly, the association's contribution in the framework of the joint marketing depended to a considerable extent on the personal contribution of the president and the members of its governing board. Specifically, the annual report of 2006-2007 titled "Farmers Federation needs everyone's contribution ", analyzing the marketing and the amount of products sold, based on the collected data, included each president's personal contribution.

The quantity of products sold as association and the contribution of each association's president for season I and season II during 2006, excluding the open field products (in kg):

Association	**First season 2006**			**Second season 2006**			**Total kg I + II**
	April	**May**	**June**	**October**	**November**	**December**	
F2 (79 farmers)	83405	147960	57981	69278	80265	74451	513340
President's contribution	7545	6979	6626	6668	6243	3973	38034
F1 (22 farmers)	408	1999	0	11302	21617	14778	50104
President's contribution	135	-	-	4771	4428	4555	13889

F3 (5 farmers) **(president has no greenhouse)**	4067 -	6102 -	1303 -	0 -	930 -	1085 -	13487 -
Federation (106 farmers)	87880	156061	59284	80580	10281 2	9031 4	57693 1
President's contribution	-	-	-	-	-	2339	2339

The contribution of the new associations to marketing, due to the process of organizing and legalizing them, began later. The most active association among the new ones was Adriatik Remas with Llazi Gjata president, and K. Lika manager This association has contributed by bringing into the warehouse various field and greenhouse products such as potatoes, watermelons, tomatoes, cucumbers, nectarines, peaches, peppers, plums, etc., and it is worth mentioning the fact that Adriatik Remas was willing to bring even more field products, but because the FFM and its customers were more focused on greenhouse products, it was difficult to sell large quantities of open field products. It was the effective leadership of this association that brought it on top of the joint marketing, leaving behind two older associations, F1 and F3..

The production statistics for January-July 2007 (in kg) for the six farmer associations including the contribution of each president are as follows:

Association	January	February	March	April	May	June	July	7 month
F2	29433	2584	25746	138654	137999	54436	38088	426940

President	3164	441	1113	11409	25292	11474	0	52893
F1	4447	1321	0	0	0	601	1782	8151
President	2121	0	0	0	0	601	0	2722
F3	0	0	754	2889	2997	318	0	6958
Adriatik	0	0	2215	14660	30297	5550	1675	54397
President	(no greenhouse)	0	0	0	0	192	550	742
Progresi	0	0	1051	5893	0	0	0	6944
President			1051	5893	0	0	0	6944
Mirëqënia	0	0	0	0	0	1841	0	1841
President	0	0	0	0	0	1283	0	1283
Federation	33880	3905	29800	162096	171293	62747	41545	505266
President	831	1321	0	0	0	0	0	2152

8. ICCO PROJECT MANAGEMENT

8.1 Management and reporting manner

The Project managers, including the managers of participating associations and the main manager, met once a week, discussed on activities and reported on the work performed. During 2004, the meetings were held alternately, at each of the associations offices, in the three villages where their centers were located, and from 2005, at the FFM office in Këmishtaj, with the participation of the sales manager and the managers of the new associations that became part of the Federation. In these meetings was analyzed the implementation of project activities and the key findings were transmitted to the associations presidents and to the project headquarters, where Daniel Hess, Sonya Harnish and the project manager worked. Thus, there was a very close relationship between the main actors of the project. Besides the marketing, the other components of the project were also analyzed: advisory service, training program, input supplies and market information network. The director of the FFM, who was at the same time project manager, regularly informed the Administrative Council and reported in writing to the Members Assembly and at the meetings of the Mennonite Foundation Governing Board, which was the contract holder with ICCO. In accordance with this contract, the project headquarters reported to ICCO by regularly submitting narrative, financial reports and annual audit reports. ICCO, on its part, always sent its feedback to the reports received, considering them regular and transparent. Thus, in the letter sent by ICCO to AFFM on 24 November 2008

(Headquarters), inter alia underlines: "*We consider the reports as approved in accordance with the mutually agreed contract...We received audited financial statements 2007. Financial reports and audit are transparent.*[9]"

The letter that I received one year later on behalf of AMMF, on 24 December 2009 on the occasion of ICCO Project Closure, signed by Mr. Hans Heijs, Head of Department for Access to Basic Services, clearly emphasizes ICCO attitude: *"We are satisfied that the project has been implemented in accordance with the mutually agreed contract..."*[10] *(See Annex 7).*

The project headquarters in Lushnje *(photo no. 16)* transferred the respective fund to the bank account of each association according to the project activities, including funds for the agricultural extension service with coupons, the salary of the association manager, office expenses, etc.

The sales manager, who managed the warehouse and maintained close relationships with the supplying farmers, presented a daily marketing report on the amount sold, types of products, prices and revenues earned from the sales to the Federation, which included the 3% commission, the earnings from weighing and the fees from cardboard boxes. The Federation used regular sales

---

[9] ICCO, Utreht, 24 November 2008. Reference: 08_064293. Subject "Feedback on Reporting" . "We consider the reports as approved in accordance with the mutually agreed contract...We received audited financial statements 2007. Financial reports and audit are transparent".

[10] ICCO & Kerk in Actie: Subject "Closure of Project File", Utrecht, 14 December 2009

book, entry records, cash records and payment documents, with the FFM logo and description.

The products entering the warehouse were recorded in an Excel file in computer by date, name of the association and name of each farmer that brought the product. Next to the name was written the amount in kg, gross ALL for the farmer, 3% commission, weighing tariff, ALL kept by the FFM and in the last column was recorded the net sum in ALL for the farmer. For those farmers who were far away, from other villages and did not have means of transport, the Federation took the production on the field with its Sprinter van and on the computer was also recorded the cost of transport.

On the framework of the project management, in August 2005 a telephone line with fixed telephone number +355 37330079 was installed in Këmishtaj warehouse.

With the establishment of the Federation and its activity, the role of the rural farmer community in the project area grew very much, a fact that Gerrit affirms in his mission report of July 2005, where he writes: "*FFM is in the vanguard of agricultural developments in Albania ..*[11]"

---

[11] G. Holtland "Support mission to the FFM Lushnja District" July 2005, pg 7

Conclusions:

- *The ICCO project management scheme was implemented through association and project managers and the reporting was done in three links: at association level, at the FFM Administrative Council and at the Project Headquarters.*
- *Monitoring by the Project Headquarters*
- *Increasing the role of the rural community through the Federation and member associations*

8.2 Problems encountered during management

The project management has faced problems as well. For example, during a work analysis done by the Administrative Council and the Director of the Federation on 29 June 2006 for the period April - May, from the verification of the amount of sales and the current status of boxes in the warehouse, resulted a discrepancy in the number of cardboard boxes that were sold with those that were recorded entering the warehouse. It was clear that not all the boxes entering the warehouse were documented, which was entirely in opposition to article 4 and 10 of the Federation's internal regulations. The sales manager accepted his mistake and the Administrative Council took the necessary measures to adjust the situation. However, to my opinion the Council should take tougher measures to him for ensuring transparency and sustainability.

At a meeting on March 2006, someone from the FFM staff proposed that the farmer's commission percentage to drop from 3% to 2% and the farmers to be excluded from the income distribution scheme without benefiting. This was a destructive proposal, it did not just help, but rather harmed all the work that the Federation had done, because it excluded the farmers and reduced the income, so I immediately opposed it, as did the members of the Administrative Council and the "proposal" was not approved by the Council. The Federation's foundation were the small farmers, and this base foundation should be broader and more solid.

In January 2006, showed up another problem, related to hiring an agronomist for the Federation. This problem caused a conflicting situation among the leaders of the associations who did not agree on the candidacy. The problem was getting more and more difficult, relations between the key leaders were affected and there was no cooperation to resolve the obstacle, and the situation was also reflected in the relationships between the staff and the leadership. Since no solution was found, I decided to resign and on January 15, 2006 I wrote a resignation letter from my position as director and the next day I submitted it to the Administrative Council, which did not accept it. The objection to my resignation came firstly by the representative of the F1 association, T.S., who turned the letter back by telling me to forget the resignation. Meanwhile in the next meeting, was solved by consensus the problem of appointment of the technician, assigning to this position N. S.

Let's go back to the problem of cardboard boxes. These boxes had a capacity of 8 kg and the FFM logo, which were a distinctive element of the Federation's marketing activity *(photo no. 17)*, and a market novelty for Myzeqe, so their management was of great importance. This was not easy because the activity was continuous, daily (except August, February and the first half of

March when the work fell off) and the boxes should be available to farmers every day, especially during the greenhouse season. The need for such boxes was great and it was estimated that for a year the FFM needed at least 100,000 boxes. In the first two months of the joint marketing (November - December 2004) the boxes were provided only by Agro Koni, while in 2005 and later they were mainly bought at two cardboard factories: initially at KRIKETS factory in Korça with administrator I. Qyteti and then at the cardboard factory in Durrës with administrator B. Balëza, with whom we had a very good cooperation. In New Year's SMS greeting in 2007, Mr. Balëza wrote: *"... you must know that you are doing a great job that nobody so far has done in Albania".* A cardboard box with the FFM logo had different prices, depending on the quality of the cardboard, the dimensions, the factory producing it and the year it was bought, but the price mainly ranged from 29 to 36 ALL. We also used smaller boxes of 6 kg. In 2007 the amount of boxes purchased was 86,280 pieces, equal to 2,591,730 ALL. We had a problem of bringing back the boxes with products by some farmers. Sometimes, these farmers, after getting the boxes to package their products, did not bring them into the warehouse, but went to Lushnja market claiming a better price, but in fact this was tiring, time-consuming and costly for them. It was a phenomenon that harmed both, them and the Federation. So, according to 2007 statistics, the number of boxes not brought back from farmers was 2,479.

Conclusions:

- *In teamwork, ethics, efficiency, correctness, sense of responsibility and cooperation with each other are of great importance; these are essential elements for every organization or co-operative, characterizing the members of staff and their leaders.*
- *Dealing with problems encountered during project management should be done without hesitance, finding the best solution for the entire grouping, facing them as a grouping, not as an individual.*

A regulation with 10 articles was drafted to consolidate the marketing activity and sales through the Federation. The text of this regulation is given below:

INTERNAL REGULATION OF FFM KËMISHTAJ FOR THE MARKETING

1. Every farmer who brings his agricultural products for sale into warehouse will be kept 3% commission and 100 ALL for each weight up to 300 kg.

2. Farmers who are not members of any federation association will pay the full value of the packaging boxes.

3. The boxes taken by the farmers in the warehouse should be brought back with products; for each box that is not brought back the farmer is obliged to pay its full value to the Federation office.

4. Any goods entering the warehouse will be accompanied by the relevant document.

5. For each good, its quantity, the stocking price and the selling price will be clearly recorded in writing.

6. For every service provided by the Federation will to farmers, it will create an income, needed to cover the expenses and to increase its sustainability.

7. The income generated by the Federation's marketing activity will be deposited on its bank account regularly.

8. Throughout the year, the payments to the bank will be made three times per month (every 10 days, respectively on dates 2, 12 and 22 of the month)

9. Information on the marketing activity carried out in the warehouse will be available daily at the Federation's Office in Këmishtaj.

10. The above articles of this regulation will be strictly respected.

FMF ADMINISTRATIVE COUNCIL

DIRECTOR

Këmishtaj, 03.12.2005

This regulation was a very important document for the promotion of the marketing activity of agricultural products through the Farmers Federation of Myzeqe.

At the end of January 2006, the project headquarters organized a survey to monitor ICCO project, with 12 questions for each participating association. Below are given the questions and the respective answers:

ASSOCIATION ________________________________

1. What is the real status of your association membership?

Answer: F1 37 members; F2 64 (46 are new members); F3 31 members.

2. How many percent of quotas have been collected during 2005?

Answer: F1 100%; F2 98%; F3 60%

3. How much ALL are deposited by the quotes in the bank account of the association?

Answer: Zero (the three associations); Managers keep the money cash and use them when the association has to cover its expenses.

4. How much money is accumulated from the sale of extension service coupons in 2005?

F1: 55,000 ALL + 810 EUR from the Village Mini-market Project; F3: 85,000 ALL + 810 EUR for the office construction; F2: 150,000 ALL.

5. Has the money from selling the coupons been deposited to bank?

F1 and F2: "No, they are kept in the cashbox." F3: "No, because they are used to construct a new association office."

6. How much is your bank account balance currently? F1, F2, F3: Minimum

7. When did your association's last elections take place?

Answer: F1, in December 2004; F2, in March 2004; F3, after the new registration in 2003 in Tirana Court.

8. When will be held the new elections for the board?

Answer: F1, December 2006; F2 and F3, within March 2006.

9. How is the relationship of the association's manager with the president?

> *The three presidents positively evaluate the work and the relationship with the respective manager.*

10. How does the association board operates and how is the manager's relationship with this board?

From the answers to question 10 it turns out that:

> *The F1 Association has the best board, the most active and the smallest in number, 5 members, of which 2 are female; F2 has 7 members but only 4 of them are active; F3 has only 3 active members out of 7 in total.*

11. At what level do you see your involvement in the joint marketing?

> *Answers are general. All the three associations agree that during 2006 they will work in this direction, but without specifying.*

12. Did you do the 2005 work analysis?

> *Only the F1 association had done an analysis, which was confirmed by its manager by submitting a written report, analyzed at the governing board meeting.*

For Association________________________________
(name, surname, signature)

Survey conducted by the ICCO Project Headquarters, monitored by the AMMF, 28.01.2006

Conclusions:

- *The FMF internal regulation was an important document that helped in the professional and transparent management of the funds and marketing activity.*
- *The survey of the associations showed that they should strengthen the aspect of the financial management and be more realistic regarding the joint marketing.*

9. THE CREATION AND DISTRIBUTION OF INCOME

Based on the marketing model, the FFM created some income from the services it provided to farmers. The income was created by:

- Having a 3% commission.

- Selling agricultural inputs (plasma, fertilizers).

-Transport service for farmers.

- Income from weighing.

- Income from membership fees paid by member associations.

- Income from marketing.

In addition, a small income was generated by selling boxes to farmers at a low price (3 ALL and later 5 ALL per box) to cover a part of their cost.

Income in 2007, one of the best years for FFM, is:

Income in ALL	For 9 months in 2007	October	November	December	Annual 2007
Comission	747.290	79.261	119.434	159.040	1.105.025
Weighing	253.520	20.468	33.900	40.200	348.088
Inputs	82.000	2.350	17.315	26.780	128.445
Transport	216.700	15.000	27.750	1.800	261.250
Membership fee	27.000	9.000	-	9.000	45.000
Marketing	132.492	-	-	40.140	172.632
Total	**1.459.002**	**126.079**	**198.399**	**276.960**	**2.060.440**

Realization of income from commission in 2007, per association:

ASSOCIATION	Commission's value (in ALL)	Percentage of commission as association
F2	829.083	75.0
ADRIATIK REMAS	147.588	13.4
F1	77.632	7.0
F3	28.759	2.6
PROGRESI	13.716	1.3
SHPRESA	4.444	0.4
MIRËQENIA	3.803	0.3
Total FFM	**1.105.025**	**100%**

A transport service by van was provided to those farmers who did not have the means to bring their own products to the warehouse, especially for the farmers of Remas who were located far away. Transportation income was regularly documented by the FFM driver A. Melishte, who was willing to work at any time and handed over the income, according to the fees set by the finance administrator, Mrs. L. Mihali.

In order to encourage cooperation between them and the continuity of joint sales, based also on the recommendation of the FFM consultant Gerrit, some of this income was distributed to farmers who brought their products to the warehouse. This was a completely new practice, accepted by farmers and associations and which gave a positive impact.

The income distributed was the one created by having a 3% commission from each farmer. Specifically, this income was divided into 3 parts, among 3 actors:

-the FMF, which benefited 40% of the amount collected.

-the relevant association, which benefited 15%, and the farmers who benefited 45%.

This scheme was implemented based on Decision no.25, dated 10/02/2006 of the Members Assembly.

On March 10th, 2006, the Federation made the first calculation and distribution of income from the 3% commission for the 11 month period, from April to December 2005 and from January to February 2006. It should be noted that during January and February the greenhouses were emptied and ready to plant at the first season of the new calendar year, so the amount collected by the commission for these two months was low, 37,454 ALL. While for the nine month period of 2005, was collected a total of 378,656 ALL, however, much less compared to the 9 months of 2007, when that amount was almost doubled, at 747,290 ALL.

The table below shows the commission quota collected in ALL during the first period applied, April - December 2005:

Month	April	May	June	July	August	Septe mber	Octob er	November	December
Value in ALL	46.102	26.122	15.425	23.028	21.454	5.845	18.702	100.020	121.958

It is obvious that the best months are November and December, belonging to the second cultivation and having tomato as the main product. According to the approved scheme, from the total amount of 378,656 ALL collected from April to December 2005, the Federation benefited 40% or 151,462 ALL and the rest of 227,194 ALL (60%) was distributed to 3 associations and their members who had brought production to the warehouse. The number of active farmers registered by the Federation who sold their products through it during 2005 was 98, out of which 73 members of F2, 15 members of F1 and 10 members of F3 association.

It is understandable that the more output sold through the Federation, the more revenue was accumulated by 3% commission and the more the three above mentioned actors would benefit. However, taking into account the expenses the Federation had for its staff, it would be better if the commission was higher, 4 or 5%, or that for a certain period of time all the commission was used only by the Federation to strengthen its sustainability and capacity for the future. In addition to the partial contribution of the project, a large part of staff payments were covered by the Federation, including the social and health insurance, the sales manager, the driver, the finance administrator and the technician. Only the value paid on social insurance for 2007 was 314,374 ALL. In addition to these expenses, the FFM also reimbursed the travel expenses of the Members Assembly or staff according to the distance, paid the cost of telephone communication, as well as the working Sundays, always based on the decisions taken by the Members Assembly and signed by the representatives of the three associations. For example, decision no.24, dated 23 January 2006 stipulated that the sales manager and the employees of the Federation were paid 800 ALL on Sundays, which abolished a previous decision when they were paid 1,000 ALL.

During 2006, the Federation paid monthly 439 EUR or 53,972 ALL (exchange rate: 1 € = 123 ALL), as explained below in the monthly salary plan of the Project and the FFM reimbursements for the payments of social and health insurance and other administrative costs:

No.	Description	Salaries from the Project		Payments from FFM		Social and health insurance		Net salary in ALL
		Value		Value				
		In €	In ALL	In €	In ALL	30.7%	11.2%	
11	Federation Technician	200	24.600	83,8	10.307	7.552	2.755	24.600
22	Federation Manager	200	24.600	83,8	10.307	7.552	2.755	24.600
33	Worker (driver)	-	-	161,5	19.866	4.298	1.568	14.000
44	Phone communication	-	-	40,7	5.000	-	-	-
55	Payment of Sundays	-	-	39	4.800	-	-	-
66	Refund of public transport for staff	-	-	30	3.640	-	-	-
TOTAL		400	49.200	438,8	53.972	19.402	7.078	63.200

In 2007, the Federation's contribution to the finance administrator was also added. For the position of the finance administrator, in fact, during the mission of Mr. Collado in October 2006, after the interviewing process, the committee analyzed the candidates who had submitted their CVs, and selected Taulant, a young boy from Goriçaj who had just completed the university on this profile, but two days later he resigned, so it was decided to nominate another candidate, Mrs. L. Mihali, who had a long working experience in the field of finance. The FFM total contribution to salaries, insurance and administrative costs in 2007 amounted to 1,188,026 ALL.

In order to strengthen the economic and financial sustainability of the Federation, there were made two decisions by the Members Assembly, one in March 2006 and the other more than a year later, in June 2007, to change the distribution ratio of the value created by the commission and consequently to increase the value of profit for the FFM. The texts of these decisions stated:

"The FFM Members' Assembly gathered on March 29, 2006 decided that from 2006 and onwards it will continue to implement the 3% commission scheme in marketing through the Federation and to distribute the income based on the financial year defined in statute under the scheme:

Federation benefits 50%, farmers 30% and associations 20% from the fund created by the 3% commission".[12]"

The text of the other decision:

"The FFM Assembly gathered today, on 1 June 2007, decided that the distribution of the commission collected from the marketing activity of the Federation for 2007 to be as follows:

1. *Federation 60% 2. Farm associations 15% 3. Farmers 25%.*

---

[12] Vendim nr. 30 i Asamblesë së FFM Këmishtaj, datë 29/03/2006

The above distribution will enter into force on January 2008.[13]"

The income generated by the sale of the boxes, was used again for their purchase, but was not enough to cover the cost, because the trader who bought the goods paid only a part of it, about 15-20 ALL / box, while the farmer paid only 3 ALL / box for the first three years of the joint sales. In 2008, taking into account the losses created by this price and also the cases when the farmers did not bring them back, by decision no.59, dated 12/01/2008 of the Members Assembly, the farmers had to pay 5 ALL / box. Part of the cost was reimbursed by the Project, but to find the best solution for the future, in early January 2007, the farmer leaders and the Federation staff discussed on this issue and provided three different versions:

-To include the box price in the sales price.

-To charge the entire cost of the box to the trader.

- To charge the trader 25 ALL and the farmer 5 ALL.

The president of the F1 association suggested that, since it was a delicate issue, it was better to discuss it with our main client, Ruzhdi Koni, who consequently was invited to discuss about finding the best solution for both parties. The proposal was constructive and so a solution was found with Ruzhdi, who had the quality to listen to the farmers and was willing to cooperate with them.

[13] Vendim nr. 53 i Asamblesë së FFM Këmishtaj, date 01/06/2007

10. ADVISORY SERVICE AND MARKET INFORMATION NETWORK

10.1 Advisory service - Coupon system

Since December 24, 2003, the associations were provided with written information on the organization of a coupon advisory service prepared by Professor Astrit, whereas in the first days of January 2004 we went to all three associations with the Professor and the expert M. Sula, in order to meet directly the office managers and the farmers, to explain them the process of advisory service, as well as to guide managers on the procedures to be followed.

The associations managed themselves the funds for the coupon advisory service. From the first fund for this purpose, on January 27, 2004, the Associations benefited from the Project the following amounts: F1 Association: 4,230 EUR, F2 1,133 EUR and F3 1,697 EUR. This distribution was made taking into account the total area of greenhouses by each association.

For 2004, the Federation made 88 coupons available, which were distributed during 2 seasons of the greenhouses. Their value was reimbursed by the project financed by ICCO. The first time, in February 2004, to the F1 association, after the transfer of 4,230 Euros, was given 32 coupons, which the association sold to the farmers based on 2 prices, 5,000 and 8,000 ALL, as explained below; the F3 association received 15 coupons and the F2 association received only 10. The second time, in September 2004, for the second culture to all 3 of the above mentioned associations the FFM distributed 31 coupons, respectively 10 to "F1", 6 to "F2", and 15 to "F3".

ASSOCIATION	First culture	Second Culture	TOTAL
	Coupon number	**Coupon number**	**Coupon number**
F1	32	10	42
F2	10	6	16
F3	15	15	30
TOTAL yr. 2004	**57**	**31**	**88**

Each of the three governing boards, within a week, from 29 January to 5 February 2004, announced the list of agronomist specialists who were selected for the coupon system. The board of the F1 association, which had 32 coupons available, was the first to approve the list of 4 persons: Dh. Qorri, N. Ndoni, Ll. Mile and Th. Stambolliu and a legal entity, the RAAC (Regional Agricultural Advisory Center) Fier. The same procedure was followed a few days later by the F2 governing board which approved two entities: RAAC Fier and N. Ndoni, having 10 coupons, and then the F3 association approved the list of specialists: A. Semini, Dh. Qorri, N. Ndoni, K. China, J. Naço and RAAC Fier for conducting the advisory service with 15 coupons available.

Following the approval of the three lists, at the end of February 2004, a joint training was organized with the participation of all the selected specialists, explaining them how to proceed for the successful realization of this service and the importance of this scheme. Through it, the associations should produce income and should know that in order to achieve this, they must provide quality services to their members. The selected specialists signed advisory contracts with the respective associations. From this contract a specialist benefited 15,000 ALL for the advisory service in a greenhouse with a surface of up to 1,000 m^2 and 20,000 ALL for the service in a greenhouse over 1,000m^2. The coupon was a document, owned by the association, which was sold to the interested farmers, as mentioned above, respectively for 5,000 or 8,000 ALL, according to the size of the greenhouse and in such a way the association created an income. The farmer was free to choose one of the approved agronomists to perform the service near his greenhouse at least once a week throughout the season.

Meanwhile, the F2 association was growing fast compared to the other associations, and at the beginning of 2005 it had 17 new members and all of them were farmers with greenhouses. For this reason, to the F2 Association were provided 21 coupons for the first culture in 2005, to the F1 association 11 and to the F3 association 17 coupons. So, in total, the Federation distributed 49 coupons during the first culture in greenhouses in 2005. For the second culture in 2005, the F1 association requested from the AMMF (project headquarters) and the FFM that the fund provided for coupons to be used by this association for the construction of a mini-market in Goriçaj, while the F3 requested to use the fund for coupons to continue the construction of the new office in Fierseman. After consulting the Foreign Advisor, Mr. Holtland, the budget of 2,500 EUR reserved for the second culture was distributed to all three associations in accordance with their respective requirements. The F1 and F3 were given 810 EUR each, while the F2, 880 EUR to continue the coupon extension service.

This extension service model helped not just farmers, but also their own associations to generate income from coupon sales. In total, the three associations created an income of 6088 EUR during the greenhouse season in 2004 and 2005. Below are shown 4 detailed tables with data for each association:

Provision of Income from the Extension Service, year 2004

ASSOCIATION	Culture I 2004		Culture II 2004		TOTAL 2004	
	Income		Income		Income	
	ALL	EUR	ALL	EUR	ALL	EUR
F1	196.000	1.555	50.000	397	246.000	1.952
F3	75.000	595	75.000	595	150.000	1.190
F2	50.000	397	30.000	238	80.000	635
TOTAL	**321.000**	**2.547**	**155.000**	**1.230**	**476.000**	**3.777**

Provision of Income from the Extension Service, year 2005

ASSOCIATION	Culture I 2005		Culture II 2005		TOTAL 2005	
	Income		Income		Income	
	ALL	EUR	ALL	EUR	ALL	EUR
F1	55.000	437	0	0	55.000	437
F3	85.000	675	0	0	85.000	675
F2	105.000	833	45.000	366	150.000	1.199
TOTAL	**245.000**	**1.945**	**45.000**	**366**	**290.000**	**2.311**

Total Income from the Extension Service

ASSOCIATION	2004		2005		TOTAL	
	Income		Income		Income	
	ALL	EUR	ALL	EUR	ALL	EUR
F1	246.000	1.952	55.000	437	301.000	2.389

F3	150.000	1.190	85.000	675	235.000	1.865
F2	80.000	635	150.000	1.199	230.000	1.834
TOTAL	**476.000**	**3.777**	**290.000**	**2.311**	**766.000**	**6.088**

Number of coupons distributed

ASSOCIATION	**2004**		**2005**		**Total coupons distributed**
	Culture I	**Culture II**	**Culture I**	**Culture II**	
F1	32	10	11	0	53
F3	1	15	17	0	47
F2	10	6	21	9	46
TOTAL	**57**	**31**	**49**	**9**	**146**

As seen from the tables above, the F1 association generated more income because it received more coupons, which means that it received a larger fund from the project since it had more greenhouses than the other two associations and besides this, it benefited also 810 EUR from the second greenhouse season of coupons for 2005 for the reasons explained above.

The experience of the coupon advisory service was presented at the Round Table with subject *"Exchange & Planning"*, organized by Agrinas in the town of Pogradec on July 6, 2004, focused on the extension service and farmers training. This activity was supported by ICCO, Netherlands, and attended by the Albanian partner organizations, Agrinas, Dorkas, FFM, AMMF, Diakonia Agapes and AFTD. Our team from Lushnja, me, Daniel Hess (the AMMF CEO) and Trifon Stambolliu (the Federation president), presented the experience with the coupon system, which was viewed as a new experience, and raised many discussions and debate among the participants.

At the end of 2007, the coupon advisory service activity was funded by the Spanish project. The coupons that were distributed during this period were sold to farmers at a lower price of 2.000 ALL / coupon, as the service period was shorter, 3 months (*photo no. 18*). The income of 2,000 ALL in this case was shared between the Federation and the respective association.

10.2 Market Information Network

To help farmers sell their products through the Federation, was created a Market Information Network which aimed at informing producers of agricultural products on the current prices from the wholesale market of Lushnje and Fier. The main role was given to the managers of 3 associations and to the project manager. The connection between traders and farmers in Lushnja market was done mainly through mobile phones and each of the managers received from the Federation 2 phone cards per month. Regular daily contacts were held with the director of Lushnja wholesale market, D. Kolleshi, who was one of the supporters and encouragers of the joint work of farmer associations and the FFM. The wholesale prices were comunicated daily to farmers by displaying them at a designated location in the three villages, near a cafeteria frequented by farmers, as well as orally. Also the prices were displayed in the warehouse where the production was accumulated, while the project manager also compiled a price newsletter containing detailed information on the selling prices of the products from Lushnja wholesale market, comparing them with the prices of the Farmers' Federation.

Conclusion:

- *The organization of the coupon advisory service was a new practice applied for the first time in Albania, a system that provided quality service to small farmers and a way to provide income to farmer associations.*
- *The market information network was a useful tool to inform small producers on the prices of agricultural products from the main markets and to identify the most required products in the market.*

11. IMPACT OF THE FFM MODEL

The model of FFM organization and functioning was being liked. Various groups of farmers and organizations visited the Federation in Këmishtaj; on 22 June 2004, a group of 23 farmers from Moldova, the Agroinform organization, visited the farmer associations in Këmishaj and was introduced to the organization scheme of the Federation and especially to the way of functioning of the coupon advisory service. The group headed by Agroinform Executive Director, Mrs. Aurelia Bondari, was composed of staff members and farmers from the areas where this Moldovan organization operated.

Also, as Director of FFM, I was a member of the delegation led by the Deputy Minister of Agriculture, Mr. Y. Tola, invited by the World Bank to attend a workshop organized by the WB, FAO and CEI from 24 to 27 May, 2004 in Sarajevo with subject "Agriculture, Agribusiness and Retail Sector in South East Europe", during which were discussed important topics on sanitary and phyto-sanitary standards, legal framework for exports to Western markets, trade liberalization and the need for competitive agriculture, sharing best practices from different countries such as Poland, Czech Republic, Croatia, etc. In our group of 5 people participated also Ruzhdi Koni, who represented the Sector of Agribusiness in Albania.

On November 9th, 2005, a group of 40 farmers organized by Dorkas organization, Korçë region, having T. Pikuli as team leader (with whom person I worked after a few years in Agrinet Foundation in Korça as coordinator of a Swedish project) visited the collecting and storage point for agricultural products in Këmishtaj and were introduced to the Federation marketing model. They visited as well some farms in Goriçaj and Mërtish villages. A month later, on December 9, 2005, some representatives from the Italian organization CEFA based in Elbasan visited the FFM. It is interesting how they found the address of the Federation. Mr. Dario, who came together with the specialist Asim Sula, told us they saw the tomato cardboard boxes with the FFM logo at the EUROMAX supermarket in QTU and dialed the phone numbers written in the boxes. They called me and the first president of the Federation Trifon Stambolliu. The FFM organization model was successfully implemented in the Elbasan district by establishing the Agricultural Federation of Elbasan (AFE, FBE in Albanian), which included associations from 3 communes, Shirgjan, Gostima and Belsh. Two years later, in September 2007, a group of 5 people went to Shirgjan, Elbasan, including the President of the FFM Associations and the Spanish Coordinator J. Cavero, who were introduced to the operation of FBE, focused on collecting olives from the members of its 3 associations and producing high quality olive oil, packed in dark glass bottles, of half liter and 1 liter, with the label "Olive Oil "Valmi".

Several years later, I had the chance to cooperate with the FBE and its staff. Owing to the commitment and the ability of the FBE leadership, Agron Çakalli and Asim Sula, the activity of this federation has moved forward. Today, the olive oil production continues to succeed, and not only in Shirgjan, the range of its activity has expanded by providing other services, as the Italian quality fertilizers and pesticides, and the agricultural mechanical service to its members. In FBE success, its president Agron Çakalli has his own deserved merit.

Going back to the activity and impact of the FFM Këmishtaj, we must mention its serious presentation with a large range of products with FFM logo at many Agribusiness fairs in Tirana, Fier, Lushnje and Berat; at the annual national fairs organized by AAC in Tirana, in 2006, 2007, 2008 and 2009. Regarding the participation in the fair organized from 29 September to 2 October 2006 and the FFM marketing experience, it was written an article in the "Agribusiness" newspaper.

On 17 June 2005, the Farmers' Federation was invited to join a meeting in Divjakë with the participation of the Prime Minister of that time, Mr. Fatos Nano and the member of Albanian Parliament Mrs. Arta Dade. During the presentation of the Federation at this meeting, and being in the main table with the Prime Minister, I exploited this opportunity (after informing the participants about the implementation of two projects funded by ICCO and CERAI), to raise three concerns on behalf of our farmers:

1. The issue of high prices of agricultural inputs
2. The abuse with the vegetable seeds - in the market were present non-original seeds which were sold to farmers as original, damaging the quality of the products and increasing the costs.
3. The involvement of the Farmers Federation in the project of reduced oil price.

In July 2006, invited by the President of Virginia Mennonite Board of Missions (VMBM), Mr. Loren Horst, I went to Virginia, USA, together with my wife to attend the Annual Conference of this organization in Harrisonburg and visited some agribusinesses, farms, livestock farms and agricultural sales markets in Virginia, Ohio and Pennsylvania. This visit was a special event and I had the chance to speak at the Conference about the great work that the Mennonite Foundation was doing in the Lushnja district for the socio-economic development of the rural community and for the organization of small farmers in associations and Federation. Loren with Earlene, his wife, invited us to their house for a visit, where we had a nice afternoon. After the Conference, firstly we traveled to Ohio, where a fascinating and interesting village environment appeared to us, the Amish Farmer community: men in black clothes with their characteristic hats, women and girls in long, clean and beautiful dresses, horse-drawn carts for the movement of this community that has not accepted cars, as well as electricity and modern technology in their homes, which was very impressive for us coming from Albania. A fruit-vegetable wholesale market in this region, was impactful by how the products were sold, exhibited in various packing and how the winning trader was determined; the sales manager spoke loudly and quickly, standing in a visible position by everyone; from the speed and the way of speaking it was difficult to understand his English. Traders and manufacturers stood focused in front of him.

During the one-month stay in America in 2006 and 2007, I met my AMMF colleagues with whom I had worked in Albania, who had served for many years in a remarkable way, characterized by their fair work and conduct, and welcomed everywhere with love and respect from the families of the villages where they served in Myzeqe. We also felt the same hospitality and affection, when invited to the house of Willy and Barbara Keener and to the house of the parents of Sonya Harnish, where we were introduced to the lives of these respectable families and visited the large farms they operated; the Keener farm focused in potato production, sold in restaurants under contracts and also the livestock farm of the Harnish family in Pennsylvania, having 130 cows. As we left Pennsylvania, we made a brief visit to Keith Blank, former EMM director, while in Virginia we experienced the same pleasure as guests to Willard and Melba Heatwole, Willard and Eva Ebberly and Daniel and Mary Hess house.

One of the unforgettable visits was the meeting with Glenn Zimmerly and his wife in Ohio in July 2006, many years after he left Albania. Although sick, Glenn had arranged several meetings for us. We traveled in a huge white Cadillac, while he used a wheelchair out of the car. We slept one night at his house and I used this opportunity to give him a symbolic present, a small table flag with the FFM's logo, for his great contribution to the organization of small farmers and rural development. A month after our return in Albania, we learned about Glenn passing away. His vision and work are a vibrant and precious example for everyone involved and working on social, economic and humanitarian development projects and activities of the rural community.

The visits and meetings in institutions, organizations, farms and agribusinesses in the United States have been a training, education and encouragement for the work and participation I had later in different projects, helping me in the performance and relationships with the colleagues and people in general.

A special event for the Farmers' Federation of Myzeqe to be mentioned is the visit and mission of the well-known Spanish television channel of international reputation, TVE. This visit took place from 11 to 14 October 2006 when a professional team of 4 people came to Albania together with Jose Luis Collado and accompanied by Davida Pertena, who was even a talented translator of Spanish. During the 4 days, was taken a lot of field footage, were conducted interviews in several villages, farms and the Federation's headquarters. The team also paid a visit to Lushnja local TV station TV4+, where they had a conversation with the TV director, Shkëlqim Bylykbashi. The Spanish television reportage was aired on the Spanish channel TVE nearly a month later, and was followed with great interest by farmers of Myzeqe area.

A group of farmers from Kosovo, organized by Intercooperation in Pristina, visited the FFM in October 2008. The group was headed by Mrs. Lumnije Rexhepi *(photo no. 19)*. Besides the field visits and the meetings with the farmers' associations in the region of Lushnja and Divjaka, the group was interested in seeing Vlora, and we visited the museum where Ismail Qemali declared the independence of Albania from the Ottoman invasion.

In December 2008, the AECID - the Spanish Cooperation, organized two very valuable seminars on the small farmers cooperation as an alternative to rural development in Albania. In these two day seminars held in Lushnje and Shkodra, among the topics referred by the Albanian professors such as Bahri Musabelliu, Sokrat Jani, Gori Stefi and prof. Antonio Gimenez from Spain, was even the FFM Këmishtaj marketing experience, for which I had prepared a Power Point presentation.

Conclusion:

- *Numerous visits to the FFM Këmishtaj, Lushnje from different organizations and groups of farmers and invitations to participate in various activities within the country and abroad, are an indication of the positive image and impact that the Farmers' Federation had established with its marketing model and other project components in Albania.*
- *The example of FFM Kemishtaj shows the effectiveness of farmers leadership in project management and involvement of small scale farmers.*

12. EXPANSION OF THE FEDERATION WITH NEW ASSOCIATIONS

"The Members' Assembly of the Farmers' Federation of Myzeqe, in the meeting of 16 May 2006 decided:

The Farmers' Federation of Myzeqe is open to accept other farmers' associations in Gradishtë or other communes of Lushnja district, while respecting its statute and being convinced that the federation's expansion will serve its further strengthening"[14]

This decision paved the way for the establishment and the membership of 4 new associations in the FFM, with the support of CERAI, the organization that managed the Spanish project. In 2006, within the framework of this project, were formed the initiator groups for the establishment of 3 new farmers' associations in 3 villages, Bubullimë, Gorre and Remas. The aim was to extend the geographic map of the FFM to other communes, besides Gradishtë, to increase its production potential and the role of the farmer community. In addition to these new groups of farmers, in Këmishtaj was created a group of women. To formalize the new groups in regular associations, initially were organized meetings in the respective villages, identifying their potential leaders. The first meetings took place in August 2006 with the assistance of the FFM, Spanish Coordinator Mr. Cavero, Mennonite Foundation, and with the contribution of the organizers of the farmer groups, Lili Koço in

[14] Decision no. 35 of FFM, 16 May 2006

Gorre, Todi Xhunga in Bubullimë and Llazi Gjata in Remas, who were willing and tireless until the end of the process for the establishment and registration of farmers' associations in the Court of Tirana. In the end, there were established the associations: "Progresi" in Gorre, "Mirëqënia" in Bubullimë and "Adriatik Remas".

The presidents of the two associations, "Progresi" and "Mirëqenia", were well-known producers, skilful to grow vegetables in greenhouses and who contributed individually to the joint marketing project, while the president of "Adriatik Remas" as a leader had the capability to attract a relatively large group of farmers to bring their products from the former Remas Commune and sell them through the FFM in Këmishtaj.

Also, 8 representatives of women from the villages of Sopez, Mërtish, Fierseman, Goriçaj and Këmishtaj, organized their first meeting at the end of August 2006 with the assistance of the Farmers Federation and the Mennonite Foundation and established the association "Shpresa".

Thus, four new associations were established and joined the Federation, and together with the three old associations, increased the role of the rural community in the socio-economic development processes and specifically the role of farmers' associations and women in these processes, which was not only the objective of the Spanish project, but also that of the previous Dutch project.

Due to decision no. 35 of the FFM and based on the project objective, already 7 associations were members of the Federation who had the same mission and were part of its projects. The challenge ahead of them was to overcome the old way of selling products and to embrace the new way of grouping together. The FFM expansion was also followed by the change of its logo, replacing the three-leafed logo with another, showing the federation image of growth: eight circles emerged from the leaf that symbolized the 7 member associations, and the last circle gave the idea of leaving the FFM door open for other associations in the future.

In 2007, all old and new member associations distributed a separate form for the identification and registration of all member farmers. Associations' managers gathered information from each farmer and created separate folders. This identification form contained valid information in two directions: data on the farmer and his family status, and data on the farm and their agricultural activity *(See Annex 8: Farmer Identification Form).*

During the first crop production in greenhouses, from March to May 2007 the work was not going as expected and the amount sold by the Federation was lower than planned. There were great opportunities and in order to improve the status of the marketing activity it was convened the FFM Members' Assembly on 1st June 2007, with the participation of the presidents of all the associations, their managers, Federation's staff and the Spanish coordinator J. Cavero; the situation was analyzed and the following decision was made:

DECISION

No. 52

Këmishtaj, 01/06/2007

The Presidents of the Farmers' Associations of the Farmers' Federation of Myzeqe gathered on 01.06.2007 and after discussing the improvement of the marketing situation in the Federation following the problems that appeared from April to May 2007 and for further strengthening it, they decided:

1. The Federation's staff must change the working style, being more active and finding the best solutions for improving the work towards marketing agricultural products.

2. The Federation's policy is the collection and sale of all fresh agricultural products of farmers who are members of the associations.

3. To increase the work in order to find new traders and markets as soon as possible for both domestic and export markets.

4. In addition to the Federation's warehouse, the sales will take place also to Lushnja wholesale market.

5. The wholesale prices must be displayed daily in Lushnja market and also in the Federation's warehouse.

6. Staff in cooperation with the presidents and the managers must effectuate an association-based planning for the second culture of 2007.

7. The Manager of the Federation working in the warehouse is temporarily replaced with its technician in order to work directly on the field, both with associations and with the market.

8. Farmers' associations involved in the Federation are fully engaged in the new Federation's warehouse project. The construction of the warehouse must start with the available funds as soon as possible and we are ready to get a soft loan from the AMMF to complement the budget for its full construction. Each association is also committed to contribute voluntarily to 5 days a month to the construction of the warehouse.

PRESIDENTS OF ASSOCIATIONS

Association F1 Goriçaj
Association F2 Mërtish
Association "Adriatik Remas"
Association "Progresi" Gorre "Shpresa"
Këmishtaj Association "Mirëqenia" Bubullimë

As a result of the analysis and the above decision, the work improved and the quantity accumulated and sold through the Federation increased significantly in the second semester, reaching 730 tons at the end of the year. The F2 association had again the biggest contribution, with 568 tons of vegetables, while the F1 contributed with 40 tons, the F3 with 21 and from the rest of the new associations, Adriatik Remas contributed with 101 tons. The total number of farmers who sold their products through the Federation increased and reached 140, divided as per below:

ASSOCIATION	Nr. farmers who sold through FMF
Association F1 Goriçaj	23
Association F2 Mërtish	71
Association F3 Fierseman	4
Association Adriatik Remas	29
Association Shpresa Këmishtaj	9
Association Progresi Gorre	2
Association Mirëqenia Bubullimë	2
Total FMF Farmers	**140**

Based on the results in 2007, during 2008 was made the planning of the quantity of products that each member association intended to bring into warehouse. The real production capacities were larger, but we took into consideration the farmers that brought their production into warehouse. The main product was doubtless the tomato and the cucumber of greenhouses. The planning of the main products for 2008 according to the associations' managers was:

Association / Culture	F2	Adriatik Remas	F1	Mirëqenia Bubullimë	Progresi Gorre	F3	Total amount in ton
Tomato	280	80	140	60	70	15	645
Cucumber	300	100	100	130	130	10	770
Watermelon	10	30	0	10	5	5	60

Pepper	0	5	0	25	10	0	40
Potato	0	30	0	0	0	5	35
Melon	7	5	5	7	10	0	34
Eggplant	5	5	0	10	5	0	25
Apple	3	7	5	0	0	5	20
Pumpkin	2	0	2	3	10	0	17
Peach	2	5	0	0	2	2	11
Lemon	1	2	2	0	3	0	8
Cauliflower	2	2	0	0	0	3	7
Onion	0	7	0	0	0	0	7
Leek	1	2	0	2	1	1	7
Okra	5	1	0	0	0	0	6
Bean sprouts	0	0	0	2	2	0	4
Apricot	2	2	0	0	0	0	4
Total ton	**620**	**283**	**254**	**249**	**248**	**46**	**1700**

The planned quota of 1,700 tons was not reached, but as outlined in section 7.2 "Marketing Model ...", during 2008 the Federation sold 816 tons of products in total.

Conclusions:

- *Of the three new farmers' associations, only one ("Adriatik Remas") managed to attract a group of producers into the joint marketing.*
- *Creation of "Shpresa" association increased the involvement and the role of women in the FFM project.*
- *The General Assembly of the FFM played an active role in making decisions for its expansion and improving the marketing situation.*

- *The challenge of the Federation remains the exploitation of the productive capacities of the associations.*

13. THE SPANISH PROJECT

13.1 Initiation of the Project

Two days before the opening of the marketing point at Këmishtaj, on 28 October 2004, Professor Civici called from Tirana and gave us a very good news, the Spanish project was signed by the Spanish authorities and soon would became official. The professor had kindly received Dhimitraq and me at the RSC (Rural Studies Center), led by him, during the talks about the new project. He was the initiator of this great project for the FFM and the man who realized the FFM connection with CERAI in Valencia for a four year period, from 2005 to 2008. In January 2005, Professor Jose Luis Collado, a passionate of rural development who was visiting Albania for the first time, was also distinguished for his noble feelings towards ordinary people and farmers. Collados's support for the development of the Farmers' Federation of Myzeqe has been very precious. As a person with great experience in the field of agricultural cooperative management in Spain, he also made a valuable contribution to the development of this Federation by transforming it to an effective cooperative according to the European model. The Spanish project titled "Support for the creation of a pilot agricultural cooperative in Albania" was presented on 24 April 2005 at a seminar organized at Hotel BLERIMI in Lushnja.

During the four years of the implementation of the Spanish project, four farming groups, most of them associations' leaders, went to Spain every year to learn from the experience of the producers and the operation of the agricultural cooperatives in Valencia and Salamanca regions.

13.2 Ecological Agriculture

The approach of this new project in Këmishtaj increased the chances for the consolidation and development of the production and management capacities of the Farmers' Federation of Myzeqe. In addition to the marketing component, the concept of ecological agriculture was introduced for the first time. A pilot group was set up with some of the best farmers in cultivating vegetables and fruit, who were members and mostly the leaders of the three associations. For the realization of the program in the field of ecological agriculture, in Tirana, was signed a cooperation agreement with the Organic Agriculture Association (OAA), which was directed by Lavdosh Ferruni and had a specialized staff. In this framework, were created some special parcels for the application of the organic farming methods in three villages, in Goriçaj, Mërtish and Fierseman.

The organic farmers' group was trained continuously by professors Enver Isufi, Florenc Kutrolli and the well-known Spanish experts from the Agricultural Research Institute of Valencia, as professor Jose Luis Porcuna, Ana Martinez, Marta Ribo Herrero, and others. *(photo no. 20)*. All the trainings developed for the ecological agriculture were quite interesting and attractive. Prof. Veri (Enver Isufi), very passionate in this field, emphasized the essence of the concept of ecological agriculture as an agricultural system that uses harmless materials for humans and other living things and harmonizes the economic, ecological and social requirements for today's and future generations. Highlighting the quality and the wonderful taste of the Albanian bio products, during a training with the pilot farmers' group, he made the following comparison: *"The nightingale is not valued for his flesh, but for the sound of his voice; so the Albanian producers cannot enter the big European market with large quantities of products, but with their quality."* The BOA and later Bio Adria Association brought to the shop of the expert N.Ndoni, who worked closely with the FFM in Këmishtaj, copper-based preparations and accessories such as pheromones for vineyards and apples, seeds of autochthonous varieties, etc., which were made available to the organic pilot group. An important part of the ecological program from 2005 to 2008 was also the provision of pilot farmers group with the bio-certificate issued by the certification body Albinspekt in Tirana, after the certification procedures and the relevant certification as Bio-Farmers by Bio Adria.

Part of the ecological program was also the inclusion of school students in order to induce them the love for agriculture and especially for the ecological one. In the framework of this program, at Fierseman high school yard, with a 1,200 EUR grant awarded by the Spanish project, was built an experimental greenhouse of 300 m^2, based on an agreement signed by the school principal and the Farmers' Federation for the application of ecological farming methods by the school students, assisted by federation's technician N. S. and two pilot farmers in Fierseman, Petraq and Andi. The school was also entitled to use the greenhouse for practical work based on the curriculum, but also took care of its protection and maintenance. The federation used the greenhouse for experiments with new vegetable varieties and demonstrations during some trainings.

During the visits to Spain, the group of organic farmers had the opportunity to visit many cooperatives which had a precious experience in the production of ecological products as well as research institutions in Valencia region, as the Agrarian Experimental Station in Carcaixent and the Advisory Service Center and Phyto-sanitary Control established since 1924 and focused on research to replace the chemical products with the biological one in the cultivation of agricultural products and plant protection.

With the assistance of OAA and Bio Adria specialists, Professor Veri, Florenc Kutrolli and Veis Selami, the Organic Farmers Group visited some bio farms in different districts of the country during 2008, which had achieved very good results: in Luz të Madh in Kavaja was visited the farm of Njazi Xhani, who was known for his passion of cultivating fruit trees with organic farming methods, especially the Pink Lady apple; in Sukth was visited E. Aliu's farm, known for the production of grapes, brandy grapes and bio wines; in Marikaj, Tirana, the group visited the agro-tourist farm, with a beautiful hilly landscape on the shore of a small lake owned by Hajdar Kuçi, who had build a complex attractive farm where you could find not only organic vegetables but grapes, olives, a small wine cellar, a livestock section with different animals, as well as the appropriate place in front of the lake where you could enjoy a lunch with bio products.

As for the training, even for the sales of the products produced by them, the pilot group collaborated with the Organic Agriculture Association to sell them to the markets of Tirana as well as to the

RSC. The land area in which 2 members of the F3 association cultivated organic cabbage and cauliflower was 3,000 m^2, in Goriçaj was cultivated organic pepper and cabbage in a surface of 2,000 m^2 and in Mërtish 1,600 m^2 (of which 600 m^2 greenhouse and 1,000 m^2 in open field) was cultivated tomato and iceberg cabbage by organic methods. So, the Farmers' Federation had a total of 6,600 m^2 for the cultivation of 5 organic vegetables *(photo no. 21)* plus 2000 m^2 of vineyard and 1 ha of apples. The sales of these organic products was in fact a problem, because the market in 2005-2006 did not recognize them as such, in terms of value and price, and despite the efforts made, the quantity sold was small and only accepted by EUROMAX in QTU, where there was also an OAA stand, while most of them were sold as conventional products.

Conclusion:

- *The ecological agriculture and preservation of autochthonous varieties thanks to the Spanish Project became one of the priorities of the Farmers' Federation with the creation of a pilot group and the production of pure bio products.*
-

13.3 Technical assistance service and agricultural inputs

Another priority of the Spanish project was the provision of technical assistance service to farmers who were members of the FFM associations. This service was free and started immediately with the hiring of a technician from the Federation, who visited the greenhouses and fruit trees of members of the three associations in some villages where they had their own farms.

In addition to this service, the Federation since 2005 also started the service of selling agricultural inputs to farmers. The company Agrokoni was supplying the Federation with quality tissues and plasmas for greenhouses, vegetable seeds, herbaceous trees, etc., which were sold at a convenient price to the farmers, while the FFM purchased the organic fertilizer AIBIOS from AIBA company by the end of 2007 and during 2008. The purchase price ranged between 12,000 - 13,000 ALL / ton and was sold to farmers by 14,500 to 15,000 ALL / ton, thus creating an income for the Federation that was necessary for the expenses and reimbursement of the technician's salary that covered the service.

13.4 Technical assistance of Spanish experts

Several months after the initiation of the project, Mr. Jose L. Collado brought in Albania a young boy, 28 years old, Jorge Cavero, who would be the Spanish project coordinator and would work until the end of this project at the FFM in Këmishtaj. Professor Jose Collado assigned me the task to be close to Mr. Cavero and to help him find an appartment in Lushnja city, and I took this task seriously. However, although I found some good options for appartments and Mr. Cavero accompanied by me, saw the options for his accommodation in Lushnja, he didn't show any interest and decided to stay in Këmishtaj village, at the house of a lady who was president of the village women association, one of the 7 associations involved in the Spanish Project. His decision to stay in the house of one of the associations' presidents for about

3 years, was also seen as not a good step, but as an action that could influence his attitudes during the project implementation...

In the second mission that Mr. Collado undertook in Albania during October 2005, he came with two very good Spanish experts, excellent professionals, one of them in the field of ecological agriculture (Mr. J. Luis Porcuna), and the other in the field of marketing and management of farmers' cooperatives (Mr. Andres Garcia). The Project's strategy was clear, to create a sustainable market chain, from producer to consumer, from farm to market by implementing the farm and market management rules.

Initially, together with professor Collado and the Spanish experts we went to the EUROMAX supermarket at QTU, where we visited the Fruit and Vegetable Department and made special meetings with Mr. Michel Brulé, Director, and Mr. A.S., Head of this Department. Mr. Brulé stressed out the fact that they accept only quality products with good packaging. He suggested the Farmers' Federation to buy simple manual equipments in Italy to prepare small plastic packaging of 1 kg of vegetables. Also on 21 October, the Federation organized a visit of 3 farmers groups from all the three associations (F1, F2, F3) to EUROMAX to see closely the quality, selection, packaging and the way of displaying fruit and vegetables. The sales line with EUROMAX functioned for several months until one day they turned back a 2 tons of bio products (tomatoes, cauliflowers, lemons), under the pressure of the producers of Ndroq area, who, of course, were very interested in selling their products to EUROMAX.

Following the mission of October 2005, field visits were organized at three farms with ecological parcels: at Trifon Stambolliu's farm in Goriçaj, which cultivated peppers and cabbages, at Stavri Gjini's farm in Mërtish, which cultivated organic grapes using also pheromones, as well as 1000 m^2 of iceberg lettuce, and at the joint farm of Andi Semini and Petraq Çapuni, who had planted cabbages and cauliflowers. In addition to the technical assistance received from the Albanian experts Enver Isufi, Florenc Kutrolli and Veis Selami, the aforementioned group was also assisted by Professor Luis Porcuna, one of the best ecological agricultural experts in Spain.

The mission of October 2005 was closed with a Roundtable, with the participation of all the actors involved in the process, where besides the Spanish experts and the staff of the Federation, participated farmers' leaders from the associations and groups of Lushnja district, field producers of flat and hilly area, professors and well-known Albanian experts in the field of rural and agricultural development Eva Dashi, Adrian Civici, Ledia Boshnjaku, Enver Isufi, Lavdosh Ferruni, Dritan Kolleshi, representatives of the Mennonite Foundation, Ministry of Agriculture, Directorate of Agriculture, Local Government and Commerce. In this open discussion was analyzed the functioning of farmers' associations in Lushnja region (Gradishtë, Krutje, Ballagat and Hysgjokaj), the role of leadership, and the involvement of small producers in marketing through associations, bringing the FFM example and its links with the supermarkets and other markets as one of the challenges of the new reality in Albanian agriculture.

The technical assistance of Spanish experts Andres Garcia, Naço, Luis Porcuna, Valentin Diaz etc. towards the implementation of the project by the Federation and farmers' associations in general and in particular for the production of ecological products, the marketing and sales aspects of agricultural products, continued by visits and by other missions throughout the project. In his mission in mid-April 2008, Mr. Valentin Diaz emphasized the commercialization of the Federation, to become known with its own products in the local and regional markets. During this mission, was also made a cost and profit estimate for cucumbers and tomatoes, that represented the two main crops of greenhouses:

a) Cucumber

Cost per 1,000 m^2 greenhouse: 244,600 ALL

Average yield during the first greenhouse season: 170 kv/1,000 m^2 = 17 kg/m2 (average figure that ranged from 10 – 30 kg/m^2).

Costs for the cucumber: 14.3 ALL/kg. Profit: 15.5 lek/kg.

b) Tomatoes

Cost per 1,000 m^2 greenhouse: 264,000 ALL

Average yield: 120 kv/1,000 m^2 = 12 kg/m^2 tomatoes (figure that ranged from 6 – 15 kg/m^2).

Costs for the tomatoes: 22 ALL/kg. Profit: 18 ALL/kg.

In an email that Valentin sent from Valencia he wrote me:

"I'm trying to offer the Federation products, and I need to know if you think is possible to offer tomatoes, thinking a truck per week (22.000 kg). Obviously starting with one and after evaluation, is probably to have an order per week..".[15]

I distributed the email translated into Albanian to Federation's leaders, asking them to present their real potentials for tomatoes and cucumbers according to their respective associations. The information provided by managers and association leaders showed that the Federation could not regularly provide the required amount for these products and the proposal was not undertaken.

Conclusions:

- *The technical assistance of the Spanish experts in the project area was a valuable contribution to the knowledge and training of farmers in Myzeqe area.*
- *The roundtables and workshops developed by FFM Këmishtaj with foreign and Albanian experts were a genuine training course for leaders of farm associations in*

---

[15] Email from Valentin Diaz, 23/04/2008 : Hi Andon, How are you? Before anything thanks a lot for your information during my last visit to Federation. I'm trying to offer the Federation products, and I need to know if you think is possible to offer tomatoes, thinking a truck per week (22.000 kg). Obviously starting with one and after evaluation, is probably to have an order per week. Could you ask for me farmers what the production calendar is, and what is in your opinion a minimum price to make our cost here? Is very important to know the calendar during all year, for tomatoes and cucumbers. Thanks a lot and I'm waiting your answer. Best regards, Valentin Diaz Pérez. Márgenes Productividad, S.L.

Lushnja region, in terms of commercial management.

- *The possibility given to FFM for the opportunity to export 22 tons of tomatoes per week to Spain, showed that the Federation was not yet ready to meet such export orders and there was still much work to be done, first of all by leaders of associations, who should be more involved and focused on joint marketing issues.*

14. VISITS TO SPAIN AND THE WAREHOUSE PROJECT

These visits were a real training experience for the participants, who were leaders of farmers' associations and FFM staff members, as well as RSC representatives. The four visits had a rich program, filled with theoretical meetings and trainings, visits to cooperatives and scientific research institutions in the field of agriculture, not lacking the educational, cultural and rural aspect as well. All these were made possible by the special contribution of two professors, J. Luis Collado and Adrian Civici.

The conclusions of the visits to Spain were presented in a special report to the Members' Assembly at its next meeting, which was drafted in Albanian and English, providing also the relevant tasks.

The first visit took place from February 28 to March 6, 2005, when the implementation of the Spanish project had just begun, a visit which is described in the report titled "European Vision for the Functioning of Co-operatives in Spain", which I presented in the meeting of FFM Members' Assembly.

Besides visiting several cooperatives and institutions of the agricultural sector in Valencia region, one of the most impressive events was waching live a football match between two of the most famous Spanish teams, Valencia - Real Madrid, on March 5, 2005. It was awesome to be in a stadium full of 55,000 people and watch the brilliant game of the Brazilian national team players, Ronaldo and Roberto Carlos, and other stars like Figo and Beckham. The match ended in a draw 1 - 1 with Ronaldo's goal.

To watch a live game was surreal and that's a dream coming true when Luis Collado, surprisingly while we were having lunch, gave us the tickets on 5 March.

The second visit took place from 2 to 9 April 2006 with the participation of 11 persons, of whom 9 represented the 3 associations and the Farmers' Federation. After this visit I prepared a report titled "A Project, a Model and ... Trust" in which, describing the visits to cooperatives and institutions of the agricultural sector in Valencia region, I emphasized the importance of reliance to one another, a fundamental concept for having an efficient, stable cooperative, and that should also characterize the relationship between the associations' leaders united in a Federation. It was necessary for them to trust one another and to support each other, considering that the three associations belonged to different rural communities. Their behavior as a leader reflected the degree of co-operation among small farmers, from which it ultimately depended the success of the project, which required a culture of trust to a higher degree.

Let us stop at the third visit to Valencia region, from 16 to 24 March, 2007. Leaders of the seven farmers' associations and FFM members were part of the group of 12 people. Throughout our visit, we noted the concern of J. Luis Collados to fulfill all its objectives. In addition to visits and meetings in agricultural cooperatives and marketing points, a special component of this visit was participation in the great week of *"Las Fallas Festival"*. The official opening of the celebrations on March 17 was wonderful, thousands of Spanish citizens marched with their popular costumes on the streets of Valencia. Many tourists from different European countries filled the hotels of Valencia to participate in this unique event and this was the reason that our group was accommodated not in Valencia but in the ancient town of Requena. The Falles Festival reached its peak on March 19th, the San Jose's day, a brilliant evening with thousands of colorful fireworks and the burning of all the big "fallas", except for one that was chosen as the best and would be kept in the special museum of the winning "fallas".

It was magnificent for the Albanian group to visit the Requena medieval museum. Large caves with giant pottery vessels for keeping olive oil, wine, wheat and other foods (each of them with a capacity of 2,000-4,000 liters) were a living proof of the Arabic culture of the 8th and 9th centuries. The Arabs used these caves as shelter to protect during the wars.

14.1 Bodegas Cueva

The visits and trainings in agricultural cooperatives and marketing points began on Monday, March 19th, with a visit to Bodegas Cueva, a wine cellar where three brothers had joined their vineyards to work together. *"When we started this kind of business some people laughed at us, but Bodegas Cueva continued to work and now we produce ecological wine and have also obtained a certificate for ecological products"* says the winery lady. While serving us a delicious wine in very large cups, she showed us that although small in capacity, the winery competed successfully with big wine companies; implemented very strong hygienic rules and was testing for the first time the production of ecological champagne. The number of workers was flexible, depending on the season. They usually had 4 employees, each with a specific task, while at peak times the need for workers ranged from 10 to 14, which represented a good organization of work to be taken into account by small groups of farmers.

14.2 Rural mountain tourism "Los Centenares"

The sense of cooperation and strong trust among its 8 members impressed us most when we visited the rural mountain tourism cooperative "Los Centenares". A small co-operative where the partners had joined their financial contributions and possessed 40 hectares of mountainous land, of which 25 ha were planted with wheat and maize, 12 ha of almonds, 2 ha of alfalfa and 1 ha of ecological apples. There were employed two experts: an agronomist and a veterinarian. The lesson learned from this small but efficient cooperative is that the number of registered members of a cooperative or association does not matter, but first the trust, acceptance and support of each other for the purpose for which they are united. The mountain restaurant of this cooperative, 150 km from Valencia, served the Albanian group with its own ecological and traditional products, for which they also had a certificate.

14.3 Coarval

One day later, the Myzeqe group had the privilege of visiting a completely different company, the Coarval giant of the Spanish co-operatives, which started its activity in 1993 with great difficulties in finding clients, but thanks to the professional management and products quality, over the years it became a leader in the sales of liquid fertilizers and agricultural inputs in Spain. Having 180 member cooperatives, which belonged to 100,000 farmers, Coarval provides various services, all kinds of agricultural inputs, fuel, consulting services, and so on. Coarval's goal was to help farmers increase their productivity at a lower cost. Two hundred agronomists worked for 180 member cooperatives, all farm records were computerized. Our group also visited two of the 60 shops that Coarval had opened for farmers and their families, where they could buy qualitative tools and work clothing, industrial and agricultural products, foods, drinks, etc., at preferential prices. President of Coarval, Mr.Jose Faus Catalá welcomed the Albanian group in the training room; his advice was: "*There are three things to be successful in our path: having a project, a model to follow, and credibility to one another.*" From these three, our farmers' federation needed especially the third, credibility, implying confidence in itself, trust in one another, and supporting each other.

Mr. Catalá also stated that Coarval's annual sales were never less than 100 million EUR. He concluded his speech by saying: *"Coarval's goal is always one: to be an efficient enterprise"*. Then an ecological lunch followed, dominated by the famous Spanish pie and other products, vegetables, honey and of course the wonderful wine.

14.4 Another ecological farm

The next day the small bus of our group stood in front of a field planted with vegetables, where the onion was dominant. The farm agronomist, Mr. Pedro Ramos was waiting for us and we walked together to a parcel of leek and salad (a special salad destined for export to France). The farm owner named Jesus, had been working on bio-dynamic farming methods for 15 years and had been awarded with a certificate for ecological products. Apart from cultivating vegetables, he also had a large surface of oranges. Pedro told us that for organic farming it is very important to have a protective fence of different shrub-like trees, because it protects the organic part of the farm from the rest that is not organic. That day there were15 tons of leek ready to harvest, but they did not harvest it because of the very low price that did not cover not even the labor force - a phenomenon that our farmers often complain about. Another interesting thing on this farm was the traditional Valencia tomato that we saw in the greenhouse. Pedro Ramos explained the practice of biological process that was used inside the greenhouse with useful insects *Encarsia Formosa* and *Eretmocerus Mundus* to protect the tomato and he added: *"bio dynamics first of all is, ecology"*.

14.5 Advisory and Phyto-sanitary Service Center

The next station for the Albanian team was a visit to the Advisory and Phyto-sanitary Service Center. The group that visited a year ago this scientific center was interviewed by a CERAI reporter, while the representatives of 4 new FFM associations (Adriatik Remas, Progresi Gorre, Shpresa Këmishtaj and Mirëqenia Bubullimë) visited the Center's laboratories and became acquainted with the motto of their work: "*Change the course from a chemical war that leads to a dead end, to a biological war, which is not an easy road anyway*".

14.6 COFRUDECA – Cooperative in Belgida

On March 22nd 2007, the group visited the 25-year "Cofrudeca" Cooperative in Belgida, very successful, with 700 members in 10 villages. Forty of them, or 5.7% of the membership, were involved in the production of ecological products - vegetables, grapes and olive oil. The annual production of the cooperative reached 12 million kg (12,000 tons) and the revenues were 6-7 million Euro / year, while the volume of ecological products reached 1.5 million kg. Members brought all production to cooperatives - a very important rule that was strictly enforced. Unfortunately, not the same for the majority of our farmers who are members of federation's associations. "Cofrudeca" is a model that must be followed in terms of organization and functioning; there are 6 departments:

- Fruit and vegetable department.

- Oil olive department.

- Wine production department.

- Sales department and store network.

- Micro loans department.

- Fertilizers department.

Expert work (agronomist): As the number of farmers in the cooperative was very large, the specialist worked with several pilot parcels and reported the results of his work to all the members by written materials and through training and visits.

Distribution of income: from the total sum are deducted the expenses (administrative expenses, packaging, etc.) and then two installments are given to the farmers. The cooperative holds 10 to 15% of the profit, the rest goes to the farmers. The share for the administrative expenses is 5 cents / kg.

14.7 Perello

Perello Cooperative that we visited on 23 March 2007, has 60 ha of land (owned by 100 farmers) of which 50 ha are greenhouses and 10 ha are open field. It can be taken as a model for our farmers, by size and organization. In the large warehouse of the co-operative there are installed several equipments: for washing, selecting, packaging the products, etc. Perello had gained the ISO certificate, hygiene was at a high level, all workers wore uniform and gloves. Packaged products were sold to MERCADONA supermarket network and were exported to various European countries, such as France, the Netherlands, Ireland, etc. The farm income was calculated at 9 EUR / m2. Quality is money; it was not accepted more than 5% of any product below the accepted standard, and if 25% of the quantity of products did not meet the standards, then a lower price was applied and when the percentage was above this figure, the entire output was returned to the farmer. In the boxes of products was not written the name of the farmer but a number that belonged to him. The farmer was autonomous, but it was the cooperative that decided what he should plant, certainly based on a detailed plan, and at this point, the agronomist's role in collaboration with the agricultural economist is of primary importance.

After the third visit to Valencia region, it was the responsibility of the group's members, associations' leaders, and the FFM staff to reflect and further improve their work, each in his own position.

Conclusions after the visit:

- *Strengthening the credibility and cooperation with each other should be a priority for associations' leaders.*
- *Each cooperative visited was paying particular attention to the issue of being an agribusiness enterprise with the utmost effectiveness, to come up with profit rather than loss, covering all administrative and other costs. The FFM's objective: to be an efficient agribusiness enterprise.*
- *FFM must make effective use of its production capacities and above all, must increase the volume of sales of agricultural products; aiming at 1,000 tons by the end of 2007, estimating that the FFM's production capacity for fruit and vegetables is about 5,000 tons per year.*
- *In every cooperative that we visited, special attention was paid to ecological agriculture as a good option for consumers and this experience is a good chance and encouragement for the FFM members organic pilot group, who have visited these cooperatives in the region and Valencia.*

- *FFM should strictly implement the rules established by the Assembly which should extend to the member farmers.*
- *The Administrative Council should operate actively and exercise its powers better.*

In April 2008, another group of FFM carried out its 4th visit to Spain, but this time not in Valencia region, but in the northwest part, in Salamanca, and the program of this visit was focused not in fruit-vegetables but in another direction, the dairy sector, with the goal of serving the objectives of the dairy project, a new FFM activity with a lot of challenges. During this one week visit (March 31st - April 6th), were visited livestock farms and cooperatives, two dairy and meat processing factories, as well as milk analysis laboratories. In addition to the visit to Salamanca, where the group was accommodated at Hotel Amefa, the beautiful city of Madrid was also visited thanks to an error on the plane ticket schedule, which did not depart at 5:00 am to Rome but late in the evening.

14.8 Warehouse project

One of the biggest contributions of the Spanish project was financing the construction of a new warehouse. To determine the place where the warehouse would be built and to negotiate the purchase of the site, the FFM Administrative Council created a 5-member committee consisting of the president, the Federation's chairman, president of the F1, president of the F3 associations and the manager of the F2 association. There were two potential options: buying an old CTS building in Këmishtaj, or buying land with sufficient storage and parking space. While the warehouse where the Federation exercised its activity until then, was not in sale. The problem that came up with the first option to buy the old CTS building was this building had 5 owners, of which only 3 of them wanted to sell their part, the other two didn't want. After the meeting of January 2006, on 6 February the commission effectuated another meeting with the representative of the owners Rr. L., who said they required 11 million ALL for the part they wanted to sell, of 4,000 m^2 land along with an almost demolished old building that required thoroughgoing repair.

Despite the negotiations to reduce the price, nothing was achieved. So remained the second option, the one of land acquisition, for which there were 2 bids, a land of 1 ha, with the price of 10 mln ALL, and the other offer of land with a surface of 4,000 m2 with the price of 4 mln ALL. The second was more acceptable for the conditions of the Federation and thanks to the loan received from the Mennonite Foundation, this land surface was purchased in an appropriate geographic position along the road "Klubi i Shoferit", in Këmishtaj.

February 2nd, 2008 marks the beginning of the construction of the new warehouse, attended also by the president of CERAI, prof. V. G. Ramón, who had come from Valencia. The implementation of this project was carried out by the construction company with President Mr. P. Bano and lasted a few months. The cost to secure the building permit from Lushnje Municipality was high, as were paid the high municipal taxes and a number of reports were prepared. There was no fiscal facility for farmers' associations, and it was a cost that was not included in the project budget but which ultimately affected the repayment of the construction company.

During its construction, at the beginning of May 2008, the Ambassador of Spain in Albania, Mr. Manuel Montobbio, accompanied by the representative of AECID for the Balkans, Mrs. Cristina G. Hernandez, General Coordinator, paid a visit to Këmishtaj. They inspected the work and held a very friendly meeting with all the leaders of the associations and the staff of the Federation *(photo no. 22)*. Mr. Montobbio talked about the support and contribution of the Spanish government to rural and agricultural development in Albania and other parts of the world, and praised the fact that the farmers of the Myzeqe area were organized into associations and established their own Federation, as well as the work done by our farmers to produce organic products that were exhibited on this occasion.

The construction of the new warehouse was completed at the end of 2008 and in January 2009 the FFM activity moved there *(photo no. 23)*.

15. EXPANSION OF FMF ACTIVITY IN THE MILK SECTOR

15.1 Benefits from the milk project

In 2008, the FFM Këmishtaj widened its activity by starting milk collection from farmers. Since 2007, there was done a great preparatory work with farmers from different villages where FFM operated to improve the hygiene conditions of keeping the cows and milk collection through meetings and seminars organized by CERAI and FFM. In order to increase the effectiveness of these meetings and to familiarize farmers with the acceptable standards of the milk parameters, were invited the specialists of the ADAMA association in Tirana. Jani Mavromati, lecturer at Agricultural University of Tirana, has contributed to increasing the knowledge of farmers in project area too. Also the visit to Salamanca and Zamora in Spain, of a FFM group in April 2008, was in the context of this project. The group members, members and leaders of farmers' associations, saw some livestock farms with cows, sheep and goats, as well as many successful dairy and meat processing businesses with advanced technology.

The collection and sale of farmers' milk started in three villages: Këmishtaj, Mërtish and Sopez. The plastic bins and buckets were removed from use and were replaced with stainless steel cans with the FFM logo for milk collection *(photo no. 24)*. The number of cows from which milk was provided by October of this year exceeded 100, which included about 90 farmer families, while the accumulated daily amount was about 2,000 liters, a quantity that fulfilled the quality parameters in terms of bacterial level , fat, proteins, somatic cells, acidity, etc. There were some collection points in the villages of the area, where small containers brought from Spain were used for the collection of milk. The staff was increased, and a hard work was done by FFM's veterinarian, A. Maka and the Spanish manager Sergio V. Mulas.

As a conclusion, farmers' benefits from this project can be summarized as follows:

- *The technical knowledge on the parameters to be met by fresh milk were widened.*
- *The hygiene of keeping cows and their milking significantly improved, affecting the reduction of the microbial load.*
- *The dairy price increased as a result of the improved quality.*
- *Farmers received free technical assistance from the veterinarian of the Federation.*
- *The plastic containers were replaced with stainless steel containers for milking, storing and preserving milk.*
-

15.2 Problems with the milk project

The benefits from this project to small farmers were just one side of the coin, certainly very positive; on the other hand, the milk sales process turned into a "headache" for the Federation for three reasons:

Firstly, the Federation's customer (a processing plant that bought milk every day) was not correct with the payments made through the bank. Farmers needed the money, but payments delayed and the FFM was under the daily pressure of farmers who were not used to retarded payments through bank, that originally went to the FFM account and then passed on to farmers.

Secondly, geographically and practically it was not easy to succeed in collecting and selling milk in an area where are located three of the most powerful milk collection and processing companies in Albania, such as Lufra, Soal and Ajka.

Thirdly, the Federation was so far focused only on a single direction: in the collection and sale of fresh fruits and vegetables, and was creating its own image in this sector, facing various challenges but with determination moving towards a clear path. While at that moment the Federation was also operating in a new direction, on livestock which was not easy. Problems began to accumulate, the amount of unpaid money was growing by passing 4 million ALL, while the small farmers were attracted by the large dairy companies, AJKA, SOAL, LUFRA, ERZENI.

Foreign business experience shows that there have been cases when powerful companies have attempted to increase the number of services or products and have failed, although they may have spent large sums of money to advertise the new product. Referring to the 12th law about the extension of the business in the "22 Immutable Laws of Marketing " by Al Ries and Jack Trout, it is explained that powerful companies such as IBM, 7 UP, Sears etc., which have expanded their business with new products have failed, losing a part of the market they possess and also it refers to the fact that if you want to be successful, you should reduce the number of products, not expand it. *"In short, the authors say, line extension means to take the brand of a successful product (e.g. A-1 sauce for steak) and to place it in a new product (e.g. A-1 sauce for chicken meat) ..... Despite the advertising budget .., A-1 for chicken meat failed".*[16]

So, the Modest Farmers' Federation, with the new line of milk collection, despite many good sides, as explained above, had problems in the field of milk management and sales, despite the great efforts made by the staff and the veterinarian's dedication, the president of the Federation and the manager, Sergio Mulas. The FFM had become and was becoming known only for farm products in greenhouses and of course their sale, while in the dairy sector it was unable to cope with the strong competition of the three milk collection and processing companies, all of which had their plants just a few miles away from Këmishtaj.

---

[16] "22 Immutable Laws of Marketing" Al Ries, Jack Trout, pg.105 - Albanian language.

16. FUTURE OF THE FFM AFTER THE PROJECT

It was the last year of the project, so at the September meeting in 2008, the Board of Directors of the AMMF, which had approved two soft grants, but relatively large loans to the FFM, one for the purchase of the land where the warehouse was going to be built and the other as part of the contribution for its construction, for which 75% of the budget was financed by the Spanish Project, I presented an information letter on the real situation of the FFM entitled *"What will be the Farmers' Federation Strategy in 2009?"* The summary of this letter was:

2008 was the last year of the project, so in September of that year, in the following meeting of AMMF Board of Directors, which had approved two soft loans, but relatively large ones to FFM, one for the purchase of the land where the warehouse was going to be built and the other as part of the contribution for its construction, for which 75% of the budget was financed by the Spanish Project, I presented an information letter on the real situation of the FFM entitled:

"What will be the Farmers' Federation Strategy in 2009?"

The summary of this letter was:

The Federation of Myzeqe Farmers (FMF) in Këmishtaj is a community-based organization, engaged in two projects: "Farmers' Federation and Marketing Program" funded by ICCO and monitored by FSHMM, and "Support for the creation of a pilot agrarian cooperative in Albania" funded by the Spanish Cooperation, monitored and managed by CERAI in Valencia and QSR in Tirana until 2007.

Since the founding of the Federation four years ago, the two projects have been focused on fresh vegetable marketing activity, giving the Federation an original image in the Albanian agricultural sector. There were 3 farmer associations, F1, F2 and F3, which established the FFM in July 2004 in accordance with law no. 8788 on NPOs.

The year 2008 is the last year of support from external sources. Now that the project is moving towards its end, it is essential that the FFM to be prepared for the next year by having a clear strategy and planning.

The question is: will the FFM be capable enough after 5 years of extensive external (not just financial) support to continue normally the activity for its members or not?

To answer this question, let's briefly look at the implementation history of the project.

... Overall, a good progress has been made in all the aspects of the project, including farmers training, fruit and vegetable marketing, extension service and organic farming, and objectives have been met at a satisfactory level. The FFM has played well its role in agricultural development in the interest of its members and has often been a training center not only for its farmers, but also for many groups from Albania and Kosovo. Over the past 4 years, the income of farm families has improved (one of the main goals of the project) and a women's association ("Shpresa") has been

established in the Këmishtaj, supported by the donor in the framework of the Farmers' Federation.

The purpose of the Federation is to facilitate the participation of farmers particularly in marketing. This participation should be based on a common vision of their development goals. In 2007, were created and included in the project four new associations from different farming communities, but not at the same level of participation. "Shpresa" Association is very well involved in training and gender issues; there are also 3 of its members involved in marketing. Having a collection and marketing center in Këmishtaj, the marketing activity has improved from year to year. Statistics show that the F2 has a steady shareholding and the largest group of farmers (on average of 74 farmers per year) who sell products through the Federation. This is the main association that has been strongly supporting the project since its biginnings. "Adriatik Remas", one of the new associations, is the second group most involved in the 2007 project with 29 farmers. Although, from 2005 to 2007 there are on average 20 members of the F1 association that brought their products to the Federation's warehouse, their participation is not distributed throughout the year. This can be partly explained by the fact that they continue to preserve early trading relationships in their village. The other old association, F3, since the beginning of the project has shown a very low participation in marketing.

On the other hand, the results of the two leaders of the F3 association on the organic product growth have been very good; they have obtained a certificate from Albinspekt for organic cabbage and cauliflower. A similar certificate was also given in 2007 to the president of the F1 association for the cultivation of organic apples. It is worth mentioning that the members of the organic group were mainly the leaders of the three old associations that were supported by the Spanish project and Bio Adria association in Tirana.

Turning to the marketing activity, there are a number of benefits provided during the project period:

- *Farmers use cardboard boxes with the FMF logo to pack their products. This is a good practice that has helped to create a good image of the Federation and to export their agricultural products.*
- *Farmers do not waste time selling their own products, but the Federation takes care about it.*
- *A van, a forklift and a tomato selecting equipment are provided by the Spanish project and serve to the associations' members and also the offices are equipped with furniture.*
- *There is a constant group of farmers who sell their products through the Federation.*
- *The latest advantage is the new warehouse of 600 m2 (twice as large as the old one) plus 300 m^2 of facilities (2 floors x 150 m^2) for offices for, workers etc., which is near its completion of construction.*

The participation of farmers and the volume of products sold through the Federation:

Year	*2005*		*2006*		*2007*		*2008 (till 31 July)*	
Association	*Nr of farmers*	*Volume in tons*	*Nr of farmers*	*Volume in tons*	*Nr of farmers*	*Volume in tons*	*Nr of farmers*	*Volume in tons*
F2 Mërtish	73	228	79	585	71	568	54	236
F1 Goriçaj	12	42	22	66	23	40	5	4,4
F3 Fierseman	10	21	5	10	4	21	1	0,2
4 new associations	0	0	0	0	42	101	10	37,4
FFM	**95**	**291**	**106**	**661**	**140**	**730**	**70**	**278**

Contribution of associations from January to July 2008 (volume in kg)

Month	*F2 Mërtish*	*Adriatik Remas*	*F1 Goriçaj*	*Shpresa Këmishtaj*	*F3 Fierseman*	*Mirëqenia Bubullimë*	*Progresi Gorre*	*FFM*
January	53331	817	4391	3695	215	0	0	62449
February	12382	0	0	0	0	0	0	12382
March	5535	195	0	0	10	0	0	5740
Aprill	45960	1145	0	1236	0	0	0	48341
May	67864	10617	0	4605	0	725	0	83811

June	*30792*	*9177*	*0*	*4154*	*0*	*0*	*0*	*44123*
July	*20273*	*683*	*0*	*0*	*0*	*0*	*0*	*20956*
Total	**236137**	**22634**	***4391***	***13690***	***225***	***725***	***0***	***277802***

A simple analysis of the situation shows that some things within the Federation need to change quickly. There are several issues that require a quick response:

- *The fact, that farmers' leaders (Members' Assembly) in some cases take a passive attitude and accept what donors recommend through their missions can be considered as a weak point. One of the cases they held a passive attitude was the period of new elections in 2007. The Members' Assembly was not able to assume the responsibility of this important moment and to elect the new Administrative Council and the President of the Federation; it took months before the Spanish mission solved the problem and the presidency passed from F1 to F2, choosing President Stavri Gjini. Of course, they have to maintain friendly relations between each other, but for the sake of the Federation's mission, leaders need to change this way of thinking and take concrete action.*
- *The farmers' leaders do not fully understand the importance of the new legislative amendments in accordance with the appropriate laws. Making such changes is a serious commitment of farmers' associations and the Federation. Although it has been*

explained the need to have a new Federation structure in accordance with Law no. 8088 on agricultural cooperatives in Albania, have passed several months and this issue has not been solved despite the fact that the associations have had all the necessary documentation to operate. Only the F2 association has recently effectuated the changes in their statute and the governing board and has submitted them to Tirana Court. All other associations are still functioning as NPOs, which is no longer the appropriate structure for a real co-operative in Albania. It is necessary and important that the current situation to change within year 2008.

- *As above described, the FMF now has a good inventory of equipment and fix assets. These assets must be maintained and used with care and effectiveness in order to serve farmers equally and maintain their destination in the future. Thanks to a soft loan provided by the Mennonite Foundation, the FFM also purchased 4,000 m^2 of land where the new warehouse was built.*
- *It is necessary for the Federation to work with a clear and detailed planning in 2009, also using the marketing plan prepared by the Rural Studies Center in Tirana. This planning should include revenue forecasts generated by marketing services (transport, commission, weighing and agricultural inputs) as well as all the necessary administrative expenses.*
- *In early June, following a CERAI mission, many changes within the Federation staff have taken place by reducing it, in order to improve the situation. Now the*

appointment of a new skilled manager is an emergency for the FMF.

The forecast is that there are opportunities and guarantees for continuity of marketing activity by the FMF next year and beyond, but in addition to the gained experience it is necessary a full transformation of the Federation into a new organization that will be able to adapt to the free market economy rules, in order to have a real agricultural cooperative, which is the main goal of the project.

In order to be a successful cooperative this new organization should have a clear and competitive strategy and should strengthen the existing ways of collaboration with those farmers who support the Federation and the cooperation.

The potential of active farmers is very high; they possess about 10 ha of greenhouses and grow large amounts of vegetables in open field. The main crops are tomato, cucumber, eggplant, peppers, okra, cabbage, cauliflower, melon, nectarine and watermelon, which reach nearly 5,000 tons per year.

The achievements of the project, besides the efforts made by the Farmers' Federation, are result of the major, unquestioned contribution of 4 prestigious organizations, ICCO, CERAI, AMMF and RSC.

Lushnje, 15 September 2008

This letter was submitted to the AMMF to inform it on the real situation of the Farmers' Federation and the fact that from that moment it should walk on its own feet. The Board of Directors of the Foundation appreciated and discussed this situation seriously and engaged to assist the Federation as it had done since its establishment.

The issues discussed in the above letter were presented in more details to the FMF Members' Assembly about two months later, in November 2008, in an 8-page report titled "On the Road to Cooperatives", highlighting two important issues as the main challenges of the Federation:

Conclusions:

The most important challenges the Farmers' Federation will face after 2008 are two:

- *The economic-financial sustainability with focus on marketing activity and the efficient use of the available tools; and*
- *The change of legal status from a non-profit organization to a cooperative.*

This report also outlined the main objectives of the FMF for 2009, among which the increase of the amount of products sold through the Federation was the first, as a key element that consequently increased the revenues, thus serving the strengthening of the economic sustainability.

17. THE NEED FOR CHANGE

In the last months of 2008, the issue of amending the Administrative Council and the transformation of the Federation from NPO into MCS, Mutual Cooperation Society, was on the agenda (meanwhile the law on SHBB, which is certainly more advanced than MCS, came out 3 and a half years later, in April 2012. Several meetings were held on this issue, attended by the representatives of farmers' associations, the FMF, the Mennonite Foundation and CERAI. At the meeting of December 5, 2008 among others were present J. L. Collado, Antonio Gimenes (from CERAI), Paul Kropf, Sonya Harnish (from Mennonite Foundation). Their presence once again showed the seriousness and importance given to the future of the Farmers' Federation of Myzeqe by these two organizations which had helped the FFM since its establishment and had invested a lot in the success of the project. A day earlier, the FFM Assembly held a meeting and based on article 18.1 of the statute, it took the decision to start the Federation's merging process and its transformation to the MCS[17], a decision which was signed by the representatives of two associations , F2 and F3, while the F1 association was not present at the meeting. This process continued for several months and to make it as transparent as possible, at the beginning of March 2009 an important agreement was signed between the three actors: Farmers' Federation, AMMF and CERAI. The text of this agreement is:

---

[17] Decision no.70 of FFM Assambly, dt 04.12.2008

AGREEMENT

Between the Farmers' Federation of Myzeqe, Mennonite Foundation and CERAI

Today, on 05/03/2009, the representatives of the three organizations, respectively FFM Këmishtaj, Mennonite Foundation and CERAI agree to sign this agreement:

1. *CERAI and the Mennonite Foundation (AMMF) are two organizations that have assisted and monitored the Farmers' Federation of Myzeqe (FFM) in Këmishtaj and its activity for more than 4 years. They initiate the preparation and signing of this agreement together with FMF Këmishtaj.*

2. *The purpose of this agreement is to support and encourage the process of merging the Federation and transforming it into an agricultural cooperative and to ensure a transparent and successful process.*

3. *The transformation process into cooperative will be conducted by the lawyer G. Ç. from Tirana and the process of merging the Federation will be directed by CPA Mimoza Sula Aliaj.*

4. *FFM Këmishtaj takes on the responsibility and obligation to ensure that the process will be transparent and fair and all the active farmers from different villages will be regularly informed.*

CERAI FFM	*Këmishtaj*	*AMMF*
(Project Director)	*(President)*	*(President)*

The AMMF interest in the success of the Federation was also demonstrated by a visit on April 19, 2008, of a group from this foundation, including the President Paul Kropf with his wife June, and the Director of Board of Members, Willard Ebberly with his wife Eva Ebberly. During this visit to Këmishtaj they were interested in FFM's marketing activity, became acquainted with the work of the staff, the new warehouse project, and organized meetings with some farmers in their greenhouses.

For six consecutive years, from 2003 to 2008, it was done a great uninterrupted work, was spent energy, was invested in people and equipment, was established a legal organizational structure for farmers, which functioned regularly and the staff had responded to different situations at any time. All this work had to be maintained. But now the whole burden was directly on the Federation and farmers' associations. Meanwhile CERAI was leaving Lushnja with a new project in Shkodra, while the Mennonite Foundation had terminated the cooperation contract with FFM and ICCO. Also, one after the other, the contracts of the federation technician, finance administrator and director had terminated. However, CERAI and AMMF did not completely interrupt the connections with the Federation, which they had supported since its foundation, while the OAA in Tirana assisted it several times during 2009.

But as of 1 January 2009, as noted above, all the main work and responsibilities were on the Federation, the Administrative Council, the FFM's President as the head of the Federation, and the sales manager who did also the job of the driver for transporting the products when needed. In addition, a girl from Këmishtaj was hired by the Federation as accountant, although inexperienced in this field, she was devoted to work. Meanwhile, the FFM had in its inventory a range of important assets and equipment to use.

It was the time to rely heavily on farmers - active members of associations who had traded their products through FFM for 5 years. For this, first of all, the Federation should be managed competently by the leaders, who had to capture this momentum, further develop the joint vegetable marketing and efficiently utilize the valuable tools available, to serve the farmers of the area.

There were not just the hardware tools; the FMF had available on its personal computers and on hard copy the documentation of the entire work performed, the daily and monthly records on the collection and sales of products, detailed data, reports, business plans that were compiled during the implementation of project and were treated in special trainings with associations' leaders and managers. Also, the Marketing Plan drafted by RSC (Rural Studies Center) in Tirana was another valuable document to be used by the FMF even after 2008, when the project was terminated. At the beginning of this 27-page study, compiled by the proffesors of AUT (Agricultural University of Tirana) Eva Dashi, Ledia Boshnjaku and Fatmir Guri, it is stated:

"The marketing plan aims to evaluate and identify the marketing and trading strategies of a business organization, and in our case, the FMF, by clearly defining its role in, local and national market, its strengths and weaknesses, as well as the opportunities and threats it offers / imposes on the environment in which it operates.

This plan outlines all the relevant objectives and the actions needed to be taken to fulfill them. Building a strategic marketing plan already constitutes an important stage of FMF's activity aiming at creating a vision oriented to all stakeholders future, further promoting unification of production techniques across all members of FFM".[18]

In addition to this plan, in order to assist specifically the FFM Këmishtaj, although already employed in Tirana, I drafted a document for planning the expected production quantity for 2009 based on the results achieved during 2007 and 2008. In a summarized way, the planning contained this assessment:

No	PRODUCT (in Tons)	Këmishtaj	Mërtish	Gorîçaj	Gradishtë	Babunjë	Remas	TOTAL (in tons)	PRICE (in ALL)	SALES (in ALL)
1	Tomatoes	130	150	80	7	8	70	445	35	15.575.000
2	Cucumber	140	1140	990	0	5	70	445	30	13.350.000
3	Peppers	1	1	0	1	1	4	8	50	400.000
4	Bamje	2	14	0	0	0	0	16	70	1.120.000

[18] "Marketing Plan for Farmers'Federation of Myzeqe" Eva Dashi, L. Boshnjaku, F. Guri, Tiranë 2008

5	Lemons	1	1	1	0	0	0	3	100	300.000
6	Eggplant	5	3	0	0	2	4	14	50	700.000
7	Nectarines	1	1	0	0	0	2	4	70	280.000
8	Melon	0	3	0	0	0	2	5	50	250.000
9	Pumpkin	2	0	0	0	0	0	2	60	120.000
10	Apple	2	1	2	5	0	0	10	35	350.000
11	Leek	0	1	1	0	1	0	3	15	45.000
12	Cauliflower	0	2	0	1	0	0	3	40	120.000
13	Potato	0	0	0	5	0	5	10	30	300.000
14	Onios	0	0	0	0	0	4	4	25	100.000
15	Lettuce	0	2	0	0	0	0	2	60	120.000
16	White beans	0	0	0	0	0	6	6	100	600.000
	TOTAL	284	319	174	19	17	167	980		33.730.000

This amount of 33,730,000 ALL, converted into EUR, was about € 270,000, which meant a substantial and sufficient sum for a normal continuation of the FMF's activity. As in the previous years of the joint marketing, the main products of the 16 agricultural products were tomato and cucumber of greenhouses, which together accounted for 85% of the sales revenue.

ICCO, in a letter sent to the Mennonite Foundation in March 2009, estimated that *"The Albanian Mennonite Mission Foundation and the FMF have acted in accordance with the joint agreement on the implementation of "Farmers' Federation and the Marketing Program" project. We consider that your findings and conclusions are realistic and correct. The Farmers' Federation is now entering a new and important stage in its growth and farmers' leaders should maintain and further advance the achievements in the key aspects of the project."*[19]

One of the biggest companies in Albania today (year 2023) for the export of fruit and vegetables in European countries is Doni Fruits Albania, and being as part of its staff in the position of Food Safety manager for about 5 years, I have seen that between Doni Fruits and the Farmers' Federation of Myzeqe, there are things in common and certainly differences. The things in common include various assets such as warehouses, offices, the forklifts for pallets, pickup and truck loading, selection machinery, truck for vegetable transportation, boxes with logo, computer equipment, etc. Of course, in general, assets of Doni Fruits Albania are at a much higher development stage, cardboard boxes are nice and diverse, depending on the product being sold, as well as the

---

[19] ICCO letter sent to AMMF, 10 March 2009

computer system is professional and has registered in its database all the supply farmers, who are equipped with an identification number and have a contract with Doni Fruits as its supplier (it happens that the names of some of the former FFM leaders are also in the electronic register of this company). The other distinct difference between these two entities is the fact that Doni Fruits Albania is a profitable business, while the FFM was a legal entity with an NPO (non-profit organization) status until September 2009 and with the membership of farmers' associations; so the internal organization is totally different.

18. CLOSING OF THE MICROCREDIT PROGRAM OF AMMF

As mentioned at the beginning, the AMMF set up a special fund to provide micro loans in order to help poor families in some villages of Myzeqe to build simple solar greenhouses, to purchase dairy cows or vegetable transport vans. Most of them were loans for greenhouses. The amount of a loan for the construction and putting into operation of a $500m^2$ greenhouse was 2.500 USD or 300.000 ALL (Albanian Lek).

At the beginning of February 2004, the Director of AFMM at that time, Daniel Hess, after I presented to him at a meeting at the project's headquarters in Lushnje the idea of farmers' associations to get microloans from the Mennonite Foundation in order to buy chemical fertilizers for their members, suggested that presidents of the associations should submit in advance a written request to the Foundation. The president and the board of AMMF accepted the request of F1 and F3 associations for micro loans to purchase agricultural inputs. In this way the associations were helped and encouraged to start an important service for their members, such as the provision of fertilizers and quality agricultural inputs, but also to create an income as an association from such services. During March 2004, the associations were given micro loans of 300,000 and 500,000 ALL each, with very favorable terms, as 6 month repayment term and 5% simple interest. Such loans were granted even later to these associations.

The closure of the micro loans program took place in a festive atmosphere on 30 September 2006 in the center of Këmishtaj, in cooperation with Coca Cola company in Tirana, which provided all the necessary assets and items with Coca Cola logo. There were invited all the people who had received loans, without exception, even those few farmers who had not yet repaid the loan for various reasons, while to all the farmers who had fulfilled the terms of the contract were awarded certificates of appreciation accompanied with a reward in ALL. The AMMF colleagues who were engaged in this program also participated, some of whom came especially from USA to Albania on this occasion *(photos no. 25 and 26).*

One of the borrowers from Goriçaj, Llazar Stavri, in his greeting speech stressed out: *"...Providing these small loans at the right time, we think had a very positive impact on agriculture, livestock and commercial business. Thus ... approximately 10 ha of greenhouses of vegetables were constructed, producing at least 15,000 tons of fresh vegetables and attaining at least a turnover of 20-25 million ALL. So, from the investments in livestock were produced 1,500 tons of milk and 50 tons of meat ... These achievements are undoubtedly dedicated to the valuable work of the Foundation members, whom we thank."*[20]

The Conclusion of the Board of Directors of the Mennonite Foundation was that the micro loan program had fully achieved

[20] "Greetings to the Albanian Mennonite Mission Foundation" Ll. Stavri, 30.09.2006

its goal of assisting poor agricultural families, urging them to work their land and improving their living in the village.

As I was at the end of this book, while reading Spartak Ngjela's book "The decline and fall of the Albanian tyranny", I was impressed by the sentence he begins his book: *"Miserable is the man who has no memories of this life and always unfortunate will be the one, who, although has many, doesn't know how to tell them".*[21] I liked these encouraging words to anyone who decides to write his memoirs and by describing in this book the events and the 10 year experience in the Myzeqe area, I feel that I have been privileged to know, learn and work with such wonderful people, leaders and rural development experts, such as Glenn Zimmerly, Willard Heatwole, Paul Kropf, Pien de Lange, Gre Kornelius, Gerrit Holtland, Jose Luis Collado, Valentin Diaz, Willy Keener, Daniel Hess and many other foreign and Albanian rural development experts, lecturers of Agricultural University of Tirana etc. who, in addition to their deep professional knowledge have had the sense of responsibility for duty, love and respect for people of rural areas, small scale farmers, and the passion to help them.

The 10 year work experience during the implementation of the socio-economic development projects in the Myzeqe area has shown and learned me the true role played by the individual and the leadership of farmers' organizations and associations,

---

[21] "The decline and fall of the Albanian tyranny" Spartak Ngjela, p. 9

confirming the important fact that true leaders are that gravitational force that attract people and the community with their vision and example for change and progress.

Photos from the Projects

Photo 1 – Board of AMMF, October 2000

Photo 2 – Inauguration of the memorial plate for Glenn & Martha Zimmerly

Photo 3 – General Assembly Meeting – F1

Photo 4- Voting for the establishment of the F1 association

Photo 5- Construction of the entrance road in Goriçaj

Photo 6- Inauguration of the village water supply

Photo 7- Advisory Service

Photo 8- Winter School with farmers – Year 2003

Photo 9- Poultry Project

Photo 10- Poultry Project

KUPON

EMRI ______________________

Eja tek QENDRA SHPRESA për rrobat.
Zgjidh çfarë ti ke nevojë.
Çdo kuadrat më poshtë përfaqëson një pikë.
Zgjidh artikujt me vlerë të barabartë me pikët që ju keni.

Mirësevini!

Photo 11- "Shpresa" Center Coupon

Photo 12- Grading machine

Photo 13- Grading of tomatoes

Photo 14- Opening of marketing point in Kemishtaj

Photo 15- Opening of marketing point

Photo 16- Project office in Lushnje

Photo 17- Products with FFM logo

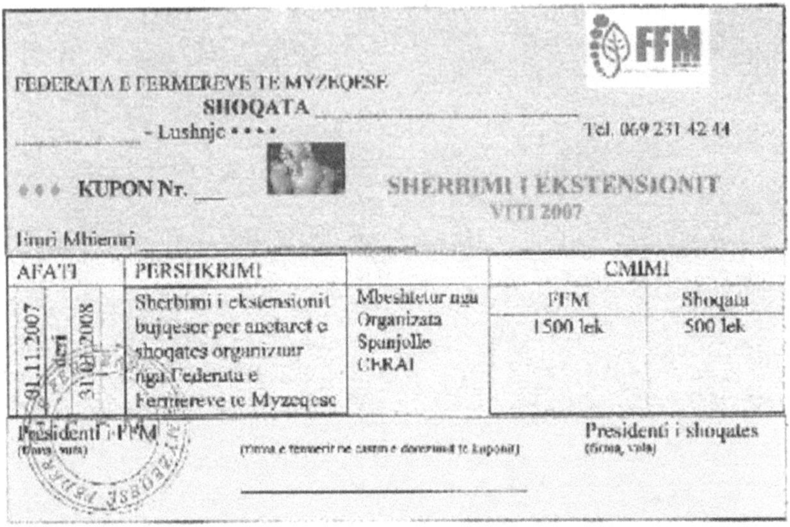

FEDERATA E FERMEREVE TE MYZEQESE
SHOQATA ____________________
____________ - Lushnje • • • •
Tel. 069 231 42 44

• • • KUPON Nr. ___

SHERBIMI I EKSTENSIONIT
VITI 2007

Emri Mbiemri ____________________

AFATI	PERSHKRIMI		CMIMI	
			FFM	Shoqata
01.11.2007 deri 31.01.2008	Sherbimi i ekstensionit bujqesor per anetaret e shoqates organizuar nga Federata e Fermereve te Myzeqese	Mbeshtetur nga Organizata Spanjolle CERAI	1500 lek	500 lek

Presidenti i FFM (firma, vula)
(Firma e fermerit ne castin e dorezimit te kuponit)
____
Presidenti i shoqates (firma, vula)

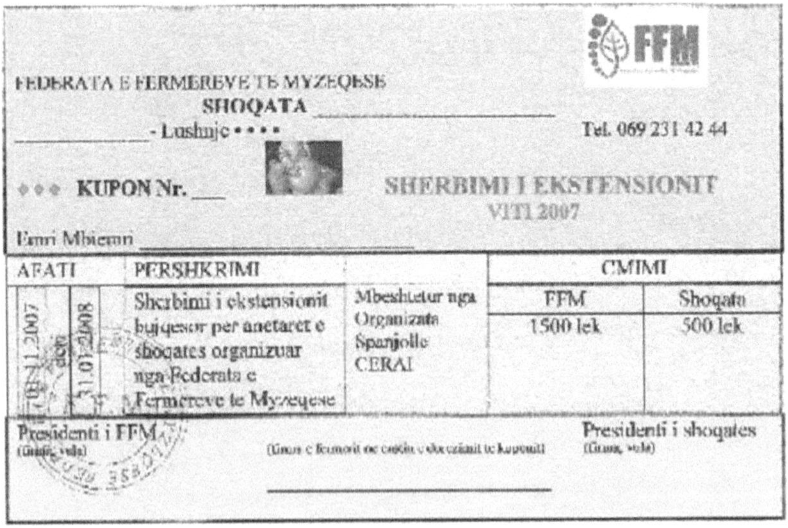

FEDERATA E FERMEREVE TE MYZEQESE
SHOQATA ____________________
____________ - Lushnje • • • •
Tel. 069 231 42 44

• • • KUPON Nr. ___

SHERBIMI I EKSTENSIONIT
VITI 2007

Emri Mbiemri ____________________

AFATI	PERSHKRIMI		CMIMI	
			FFM	Shoqata
01.11.2007 deri 31.01.2008	Sherbimi i ekstensionit bujqesor per anetaret e shoqates organizuar nga Federata e Fermereve te Myzeqese	Mbeshtetur nga Organizata Spanjolle CERAI	1500 lek	500 lek

Presidenti i FFM (firma, vula)
(Firma e fermerit ne castin e dorezimit te kuponit)
____
Presidenti i shoqates (firma, vula)

Photo 18- Model of coupons

Photo 19- Visit from Intercooperation Kosovo

Photo 20- Albanian and Spanish specialist of ecological agriculture

Photo 21- Organic Products

Photo 22- Visit of the Spanish Ambassador to FFM, May 2008

Photo 23- New warehouse

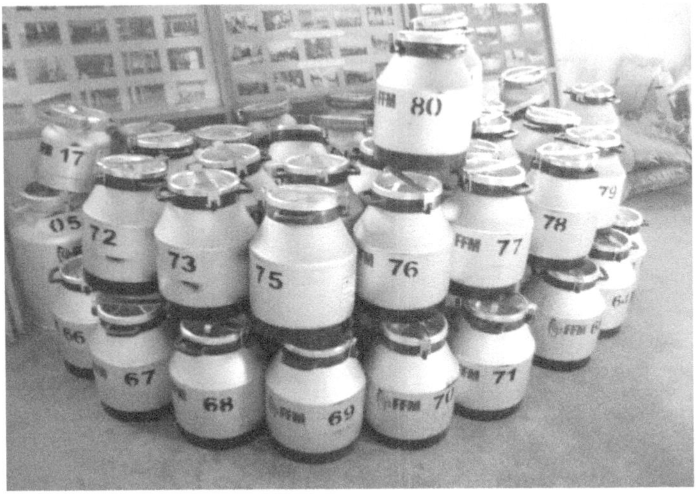

Photo 24- Milk cans

Photo 25- Closure of micro-loans program

Photo 26 – Closure of micro-loans program

Annexes

Annex 1 Operating Agreement

LUSHNJE, ALBANIA

December 28, 2000

Operating agreement between ICCO, FERGOR, AMMF, RAAC

1. For the FERGOR Projects of Drinking Water, Road Building and the Telephone System the FERGOR Board of Directors will have responsibility for getting those 3 projects underway and on schedule. The FERGOR Board must be willing to help with the exact implementation in the Gorioaj community to fulfill the requirements according to the contract with ICCO. Also, they must provide the commitment of the board, association members as well as the whole community. FERGOR is responsible to keep in close contact with ICCO through Andon Rrapokushi of AMMF with timely reports of progress.

2. Albanian Mennonite Mission Foundation (AMMF) will be responsible for the management of the Winter Farm School, the Agricultural Extension Program, and the Poultry Project to be financed by ICCO. For example, it will be the responsibility of AMMF to be in contact with RAAC to determine together the course subjects and the guest instructors as needed. There will be a separate contract between AMMF and RAAC.

3. The Projects Office and Service Center will be under the management of Andon Rrapokushi and Willard Keener of AMMF. The staff will be selected by AMMF with approval of ICCO. Staff members will be paid from ICCO funds for Service Center and Office and will be managed by AMMF.

4. RAAC will be responsible to accomplish the contract with AMMF for the Winter School and Ag. Extension Program. Zhaneta Zylaj, chief of RAAC-Fier, will be contracted to provide training for Board members of FERGOR and will come to Goricaj for the training.

FERGOR ASSOCIATION

AMMF

Glenn Zimmerly
Andon Rrapokushi
Willard Keener

ICCO
Pien de Longe

RAAC
Zhaneta Zylaj

FAX +31 (0) 30 692 56 14
+31 (0) 30 692 79 87

Annex 2: Email from Pien de Lange

Pien de Lange, 03:31 PM 1/11/01 , Betr.: Letter from Andon

Return Path: <Pdl@icco.nl>
Date: Thu, 11 Jan 2001 15:31:22 +0100
From: "Pien de Lange" <Pdl@icco.nl>
To: <anesk@abissnet.com.al>
Subject: Betr.: Letter from Andon
Content-Disposition: inline

Dear Andon,

Congratulations with the signed agreement with FERGOR, ICCO, AMMF and RAAC.

The second instalment by ICCO can be expected this month!

I agree with the decision of the FERGOR Board and Glenn.

Did you start already with the Winter School?

Best wishes,

Pien de Lange

Pien de Lange-Lunenborg
Europe Desk : ICCO
P.O. Box 151
3700 AD Zeist, The Netherlands
Tel: + 31 30 69 27 817
http://www.icco.nl
Email: pdl@icco.nl
Fax: + 31 30 69 25 614

Printed for anesk@abissnet.com.al 1

Annex 3:

Visiting small farms in Lushnje, Albania

[illegible] and I joined the Mennonite mission team in Lushnje, Albania on May 2. Almost every day there are interesting—[illegible] exciting [illegible] things taking place.

Try going to the store to buy basic groceries without knowing [illegible] word for any of the items you want. You might think that would be easy. Just push the grocery cart down the aisle and pick up what you want. The problem with that idea is that there are no "serve yourself" grocery stores in town. But with the help of a dictionary or a friend, you would soon learn the basics of living in another part of God's world.

Andon [illegible] is a 47-year-old husband, father, neighbor, Christian brother and co-worker who is fluent in Albanian, English and Greek. He also understands Italian and Russian. Andon is a great help to us as a translator. He translates for the Friday evening Bible study in our house and has translated twice when I preached in the local churches at Gorica[illegible] and Remishta[illegible].

Andon and I also work together with the farmers. Yesterday we took an extension agent on a tour of greenhouses where farmers were almost ready to transplant tomato seedlings. At one place the seedlings had been moved from a protected area to full sun in one day and showed signs of serious stress. The extension agent spoke rather strongly to the farmer, saying that it was not good for the seedlings to be treated so harshly and that the plant's first setting of fruit would be retarded. After we left the farmer I expressed my support for the agent for his speaking the truth even when it was not a pleasant task. I [illegible] that in the end, it is the truth that will set people free. With Andon's help, we [illegible] discussion on truth, with a Bible-centered testimony.

Much of our effort is in making connections and gathering important information that the small farmers need. Recently a member of the church had the unfortunate experience of losing their family cow. Just a couple days earlier we had talked with a missionary couple who had helped three poor families each buy a cow. We called the missionary, who lives about 35 miles away, and presented the need. He said he had almost enough money in his account to buy one more cow and would be happy to help.

We arranged for the missionary couple to meet the family, see the stable and hay stack, and take some pictures. They are now trying to find a suitable cow to purchase. This needy family will soon

[illegible] Hestvedt [illegible] from left [illegible] an Albanian extension agent (right) [illegible] [illegible] Albania.

have a cow again to meet their milk needs and provide some income for the household budget. The first calf is to be given back to the church for another family in need.

After the morning church service on August [illegible], Melba and I plan to leave by car for the neighboring country of Greece. Andon and his wife, Angelina, will accompany us. Angelina was born in Albania to ethnic Greek parents, so she speaks Greek as her first language. They will team up to translate for us. Angelina will translate Greek to Albanian, and Andon, Albanian to English. We will

Andon [illegible] and [illegible] Hestvedt [illegible] Lushnje, Albania.

be going to a vegetable growing area and hope to learn how the farmer's association functions, some production techniques since the climatic conditions are very much like ours in [illegible] area, and methods of marketing and distribution.

VMBM is in need of more missionaries for Albania and other parts of our world. Are you available for a [illegible] assignment? We can [illegible] being cross-cultural missionaries. If you sense God's call for this kind of service, begin conversations with a staff person at VMBM.

— [illegible] Hestvedt
Lushnje, Albania

3

Annex 4: MCC MOU 2003

A Christian resource for meeting human need

Mennonite Central Committee | Southeastern Europe Office | Zagrebačka 18
71000 Sarajevo Bosnia-Herzegovina | Tel +387 33 715 080 Fax: +387 33 613 324
Mennonite Central Committee Home Office, Akron, Pa. 17501 U.S.A.

MCC

MCC MEMORANDUM OF UNDERSTANDING FOR FINANCIAL GRANTS

Date: 21.February 2003
Partner Organization: EMM
Project Title: Farmer's Federation
Project Location: Lushnje, Albania
Project Timeline: December 2002-July 2003
MCC Contributed Amount: USD 2.000 Total Project Budget: USD 2.000

This memorandum of understanding represents an agreement on the use of project funds from MCC to EMM

It is agreed that:

1. Project funds from MCC will be used for the purposed outlined in the project description (attached). Any changes to the intended use of the funds must be approved by MCC.

2. EMM will keep receipts for money spent and make them available for inspection by MCC.

3. MCC will have the right to audit financial records related to the project
1)

4. EMM will submit reports semi-annually and a final report not more than one month following the completion of the project. All reports will follow formats either provided by MCC or mutually deemed adequate.

5. EMM will return unused funds to MCC or redirect them towards other needs following approval of MCC.

6. EMM will allow project visits by MCC.

Signed/dated: | Signed/dated:

Mennonite Central Committee | EMM J. Willard Kraus 4 March, 2003 J.W.K.

Annex 5: Dixie letter to Glenn Zimmerly

Worldwide

SINCE DIXIE 1895

Dependability

dixie canner company

BOX 1348 • 786 EAST BROAD STREET
ATHENS, GEORGIA 30603-1348 U.S.A.
706-549-1914 FAX: 706-549-0137
E-mail: sales@dixiecanner.com

Yes, you can ...

... with a Dixie Double Seamer!

CANS MAY BE:
Plastic or PET, Aluminum, Composite, Tin or Steel.

TOPS MAY BE:
Steel, Tin, Aluminum, Pop-Top, E-Z Open, LOE, Peel Off, require a can opener, Aerosol, New or Reusable ... which you may pump out the air for freshness!

www.dixiecanner.com

August 24, 1998

Check out Dixie Canner on the Internet
www.dixiecanner.com

Glenn Zimmerly
Lagja: Gafur Muco
Lushnje
ALBANIA

Dear Mr. Zimmerly:

Thank you for your inquiry on Dixie equipment. Enclosed please find our latest brochure on Dixie Double Seamers, Retort Packages, and Specialty Canning Equipment. All equipment is priced with regular equipment as shown.

Dixie Double Seamers are customized for your container. Each seamer is priced with regular equipment set to close one size can. Change parts are required to close additional can sizes within the range of the machine. Each different diameter or style of lid require a separate chuck; each different height, style or diameter can require a separate base plate. A height spacer may be required for short cans, or for use with vacuum seamers. Various profiles of stainless steel seam rolls with bushings are available at additional cost for specific applications; some can and lid configurations require the use of special seam rolls.

The Retort Brochure shows four different packages, with Package #1 including the basic retort with regular equipment as itemized. Package #2 includes minimum accessories for processing #3 (404 x 414) or smaller metal cans. Package #3 includes minimum accessories for processing larger cans and/or glass jars. Package #4 includes instruments for semi-automatic operation, and pressure and temperature recording; please note that optional equipment is required for over-pressure cooling of glass jars or larger containers. Piping, tubing, hand valves, fittings, etc. are not included for any package. Only listed items are provided.

Once you have had an opportunity to review our brochure, and have selected specific items of equipment, we would be happy to prepare a Proforma Invoice for you. Please provide electrical requirements and shipping directions for a more accurate quotation. For can double seamers, please provide diameter x height of each can to be closed.

Please do not hesitate to contact us if you should require additional information.

Sincerely,
DIXIE CANNER CO.

Cynthia J. Simmons

Cynthia J. Simmons
Office Manager

IF you wrap it, bag it, box it ... Why not can it? IT may be easier, faster, better protected IF you can IT.

Annex 6: Agreement AMMF – Farmer Associations (pg. 1)

Annex 1

AGREEMENT

BETWEEN THE ALBANIAN MENNONITE MISSION FOUNDATION AND ASSOCIATION

For the realization of the project "INCREASE OF INCOME OF FARMER FAMILIES THROUGH STRENGTHENING THE CAPACITIES OF FARMER ASSOCIATIONS AND IMPROVING THEIR ORGANIZATIONAL STRUCTURE".

This agreement is signed today, on___.___2003, in Lushnja, between the Albanian Mennonite Mission Foundation, implementator the above project, represented by Mr. Daniel Hess, Executive Director and Mr. Andon Rrapokushi, and the Association ________________represented by Mr.______________, president of the association and Mr.________________, secretary (or vice president) of the association.

The agreement aims to determine the competencies and obligations of each party, in the interest of the precise implementation of the project and the achievement of all its objectives.

The Mennonite Foundation, in accordance with the contract signed with the main donor of the project ICCO, will be the main responsible for the implementation of the project.

In this context, its functions will include, but will not be limited to:

- Opening of a special bank account and conducting all bank transactions with the project donors.
- Collaboration with the association_______________ to detail and define the timeframes for the implementation of the plan of activities foreseen by the project.
- Monitoring the plan of activities that various associations must carry out within this project and the necessary interventions to realize the required corrections.
- Monitoring and verification of indicators set to assess the realization of the project objectives by the respective association.
- Disbursement of the relevant project funds, after the implementation of the respective monitoring indicators, in the bank account of the association_______________.
- Coordination of activities among all beneficiary organizations of this project.
- Coordination of activities aimed at the creation, registration and operation of the Farmers' Organizations Union (Federation).
- Selection and coordination of the activity of local consultants that will engage in the realization of specific project requirements.
- Collection of relevant data on the performance of project objectives and compilation of periodic reports in accordance with donor requirements.

Annex 7: Winter Farm School 3

FSHMM - QRKB FERGOR

KALENDARI I SHKOLLES SE DIMRIT Nr. 3
GORIÇAJ-2002

MUAJI	JANAR	SHKURT	MARS	PRILL
Nr.Aktiviteteve	3	4	4	1
Datat e aktivitetit dhe lloji.	(17)(per serat) (24)(tmjnnn-Fier) (25)(per serat)	(1)(ne Tirane) (8)(per serat) (15)(blegtori) 22 (per serat)	1 (per serat) 8 (blegtori) (15)(per serat) 22 (Fond Men.).	Nje udhetim ne Kronet me grup fermeresh

Shenim. Disa leksione do te shoqerohen me videokaseta perkatese.

KALENDARI I PROGRAMIT TE EKSTENSIONIT BUJQESOR
GORIÇAJ – KEMISHTAJ

FAZA E PARE						
Muaji	Dhjetor	Janar	Shkurt	Mars	Prill	Maj
Nr.vizitave	1	1	4	3	2	3
Data dhe lloji i vizites	(20), ne sera.	(31),blegtori	(7),ne sera 14, ne sera 21, ne sera 28, ne sera	7, ne sera 14, ne sera 21,blegtori	4, ne sera (11) ne sera	(2),ne sera (9),blegtori (16),ne sera

FAZA E DYTE				
Muaji	Gusht	Shtator	Tetor	Nentor
Nr. i vizitave	2	2	2	1
Data dhe lloji i vizites	(15), ne sera (22),ne sera	(5),ne sera (12), ne sera	(3),ne sera (10),blegtori	7, ne sera.

EXTENSION PROG: the experience shows that the real need for visits in the greenhouses during the period of Feb. – March is more than 1 day per week. The specialist has many requests from the farmers during this period.

Annex 8: Winter Farm School 4

FSHMM QHKB – PERGOR – FITORJA – GJELBERIMI

KALENDARI I SHKOLLES SE DIMRIT Nr. 4
VITI 2003

GORIÇAJ

MUAJI	JANAR	SHKURT	MARS	PRILL
Nr.Aktiviteteve	2	1	1	1
Datat e aktivite-tit dhe lloji.	10 (per serat) 31 (per serat)	14 (blegtori)	14 FSHMM – trajnim me 3 bordet.	Udhetim pune me grup fermeresh

FIERSEMAN

MUAJI	JANAR	SHKURT	MARS	PRILL
Nr.Aktiviteteve	1	2	1	1
Datat e aktivite-tit dhe lloji.	17 (vreshtari)	7 (patatja) 21 (perime)	14 FSHMM – trajnim me 3 bordet.	Udhetim pune me grup fermeresh

MERTISH

MUAJI	JANAR	SHKURT	MARS	PRILL
Nr.Aktiviteteve	1	2	1	1
Datat e aktivite-tit dhe lloji.	24 (per serat)	7 (per serat) 28 (blegtori)	14 FSHMM – trajnim me 3 bordet.	Udhetim pune me grup fermeresh

Shenim: Disa leksione do te shoqerohen me videokaseta perkatese

FSHMM, Lushnje - JANAR, 2003

Annex 9

KONTRATE

9 Mars 2001

Kjo kontrate lidhet sot, me 9 Mars 2001, ndermjet Shoqates FERGOR - Goricaj, Lushnje te perfaqesuar nga presidenti i shoqatas Refail Thomai dhe zevendes presidenti Trifon Stambolliu dhe Shoqerise "SEMAN" SHPK - Gjirokaster te perfaqesuar nga Vangjel Goga dhe Pandeli Kuremeno, me

OBJEKT:

- Kryerja e shpimit te nje pusi hidrogjeologjik ne perputhje me projektin e paraqitur nga prof. dr. Romeo Eftimiu te shoqerise ITA Consult - Tirane.
  - Firma "SEMAN' eshte e detyruar te zbatoje projektin me perpikmeri.
  - Cdo ndryshim eventual ne varesi te prerjes faktike gjeologjike do te behet vetem me miratim te projektuesit, i cili eshte pergjegjes per projektin.

VLERA E KONTRATES

200 meter linear x 19.000 lek/m = 3.800.000 lek. Kjo vlere perfshin:

- a) Shpimin 200meter linear.
- b) TVSH 20%

MENYRA E PAGESES

1) 30% e vleres se kontrates do te jepet ne diten e fillimit te punes.
2) 60% e vleres se kontrates ne perfundim te punimeve, qe presupozon:
    a) Shpimin e pusit ne thellesine e parashikuar ne projekt.
    b) Tubimin e pusit.
    c) Zhvillimin e pusit.
    d) Pompimin dhe prurjen e parashikuar.
    e) Marrjen e analizave te ujit te pijshem.
3) 10% e vleres se kontrates do te paguhet ne fund, kur te dorezohet raporti hidrogjeologjik.

Per Shoqaten FERGOR - Goricaj

Presidenti **Refail Thomai** ____________

Zevendes presidenti **Trifon Stambolliu** ____________

Per SHPK "SEMAN"

Vangjel Goga ____________

Pandeli Kuremeno ____________

About the author

After spending many years in the secondary education systems a physics teacher, and high school superintendent, since 1999 **Andon Rrapokushi** has been involved in the rural and agricultural development sector. During this time, he has led countless development projects in rural communities across Albania – drawing on a rich experience in social impact programs. He currently works as an agricultural and food industry consultant, advising large organizations in Albania on adapting best practices. *The Power of One* is his first book.

Made in United States
Cleveland, OH
05 November 2026

47664001R00126